Tharwat F. Tadros

**Dispersion of Powders in Liquids
and Stabilization of Suspensions**

Related Titles

Tadros, T. F. (ed.)

Topics in Colloid and Interface Science

2012
Hardcover
ISBN: 978-3-527-31991-6

Tadros, T. F. (ed.)

Self-Organized Surfactant Structures

2011
Hardcover
ISBN: 978-3-527-31990-9

Tadros, T. F.

Rheology of Dispersions

Principles and Applications

2010
Hardcover
ISBN: 978-3-527-32003-5

Tadros, T. F.

Colloids in Paints

Colloids and Interface Science Vol. 6

2010
Hardcover
ISBN: 978-3-527-31466-9

Tadros, T. F. (ed.)

Colloids in Agrochemicals

Colloids and Interface Science Vol. 5

2009
Hardcover
ISBN: 978-3-527-31465-2

Tadros, T. F. (ed.)

Emulsion Science and Technology

2009
Hardcover
ISBN: 978-3-527-32525-2

Tadros, T. F. (ed.)

Colloids in Cosmetics and Personal Care

Colloids and Interface Science Vol. 4

2008
Hardcover
ISBN: 978-3-527-31464-5

Tharwat F. Tadros

Dispersion of Powders in Liquids and Stabilization of Suspensions

WILEY-VCH Verlag GmbH & Co. KGaA

The Author

Prof. Dr. Tharwat F. Tadros
89 Nash Grove Lane
Wokingham, Berkshire RG40 4HE
United Kingdom

All books published by **Wiley-VCH** are carefully produced. Nevertheless, authors, editors, and publisher do not warrant the information contained in these books, including this book, to be free of errors. Readers are advised to keep in mind that statements, data, illustrations, procedural details or other items may inadvertently be inaccurate.

Library of Congress Card No.: applied for

British Library Cataloguing-in-Publication Data
A catalogue record for this book is available from the British Library.

Bibliographic information published by the Deutsche Nationalbibliothek
The Deutsche Nationalbibliothek lists this publication in the Deutsche Nationalbibliografie; detailed bibliographic data are available on the Internet at <http://dnb.d-nb.de>.

© 2012 Wiley-VCH Verlag & Co. KGaA, Boschstr. 12, 69469 Weinheim, Germany

All rights reserved (including those of translation into other languages). No part of this book may be reproduced in any form – by photoprinting, microfilm, or any other means – nor transmitted or translated into a machine language without written permission from the publishers. Registered names, trademarks, etc. used in this book, even when not specifically marked as such, are not to be considered unprotected by law.

Cover Design   Adam-Design, Weinheim
Typesetting   Toppan Best-set Premedia Limited, Hong Kong
Printing and Binding   Markono Print Media Pte Ltd, Singapore

Print ISBN: 978-3-527-32941-0
ePDF ISBN: 978-3-527-65662-2
ePub ISBN: 978-3-527-65661-5
mobi ISBN: 978-3-527-65660-8
oBook ISBN: 978-3-527-65659-2

Contents

Preface *XIII*

1 **General Introduction** *1*
1.1 Fundamental Knowledge Required for Successful Dispersion of Powders into Liquids *1*
1.1.1 Wetting of Powder into Liquid *1*
1.1.2 Breaking of Aggregates and Agglomerates into Individual Units *8*
1.1.3 Wet Milling or Comminution *8*
1.1.4 Stabilization of the Resulting Dispersion *9*
1.1.5 Prevention of Ostwald Ripening (Crystal Growth) *9*
1.1.6 Prevention of Sedimentation and Formation of Compact Sediments (Clays) *10*
1.2 Particle Dimensions in Suspensions *11*
1.3 Concentration Range of Suspensions *11*
1.4 Outline of the Book *12*
References *16*

2 **Fundamentals of Wetting and Spreading** *17*
2.1 Introduction *17*
2.2 The Concept of the Contact Angle *18*
2.2.1 The Contact Angle *19*
2.2.2 Wetting Line – Three-Phase Line (Solid/Liquid/Vapor) *19*
2.2.3 Thermodynamic Treatment – Young's Equation *19*
2.3 Adhesion Tension *20*
2.4 Work of Adhesion W_a *22*
2.5 Work of Cohesion *22*
2.6 Calculation of Surface Tension and Contact Angle *23*
2.6.1 Good and Girifalco Approach *24*
2.6.2 Fowkes Treatment *25*
2.7 The Spreading of Liquids on Surfaces *25*
2.7.1 The Spreading Coefficient S *25*
2.8 Contact Angle Hysteresis *26*

2.8.1	Reasons for Hysteresis	28
2.8.2	Wenzel's Equation	28
	References	29

3   The Critical Surface Tension of Wetting and the Role of Surfactants in Powder Wetting   *31*

3.1	The Critical Surface Tension of Wetting	31
3.2	Theoretical Basis of the Critical Surface Tension	32
3.3	Effect of Surfactant Adsorption	33
3.4	Dynamic Processes of Adsorption and Wetting	34
3.4.1	General Theory of Adsorption Kinetics	34
3.4.2	Adsorption Kinetics from Micellar Solutions	36
3.4.3	Experimental Techniques for Studying Adsorption Kinetics	37
3.4.3.1	The Drop Volume Technique	38
3.4.3.2	Maximum Bubble Pressure Technique	39
3.5	Wetting of Powders by Liquids	42
3.5.1	Rate of Penetration of Liquids: The Rideal–Washburn Equation	43
3.5.2	Measurement of Contact Angles of Liquids and Surfactant Solutions on Powders	44
3.5.3	Assessment of Wettability of Powders	45
3.5.3.1	Sinking Time, Submersion, or Immersion Test	45
3.5.3.2	List of Wetting Agents for Hydrophobic Solids in Water	45
	References	46

4   Structure of the Solid–Liquid Interface and Electrostatic Stabilization   *49*

4.1	Structure of the Solid–Liquid Interface	49
4.1.1	Origin of Charge on Surfaces	49
4.1.1.1	Surface Ions	49
4.1.1.2	Isomorphic Substitution	50
4.2	Structure of the Electrical Double Layer	51
4.2.1	Diffuse Double Layer (Gouy and Chapman)	51
4.2.2	Stern–Grahame Model of the Double Layer	52
4.3	Distinction between Specific and Nonspecific Adsorbed Ions	52
4.4	Electrical Double-Layer Repulsion	53
4.5	van der Waals Attraction	54
4.6	Total Energy of Interaction	57
4.6.1	Deryaguin–Landau–Verwey–Overbeek Theory	57
4.7	Flocculation of Suspensions	59
4.8	Criteria for Stabilization of Dispersions with Double-Layer Interaction	62
	References	62

5   Electrokinetic Phenomena and Zeta Potential   *63*

5.1	Stern–Grahame Model of the Double Layer	67

5.2	Calculation of Zeta Potential from Particle Mobility 68
5.2.1	von Smoluchowski (Classical) Treatment 68
5.2.2	The Huckel Equation 71
5.2.3	Henry's Treatment 72
5.3	Measurement of Electrophoretic Mobility and Zeta Potential 73
5.3.1	Ultramicroscopic Technique (Microelectrophoresis) 73
5.3.2	Laser Velocimetry Technique 76
5.4	Electroacoustic Methods 78
	References 83

6	**General Classification of Dispersing Agents and Adsorption of Surfactants at the Solid/Liquid Interface** 85
6.1	Classification of Dispersing Agents 85
6.1.1	Surfactants 85
6.1.2	Anionic Surfactants 85
6.1.3	Cationic Surfactants 86
6.1.4	Amphoteric (Zwitterionic) Surfactants 86
6.1.5	Nonionic Surfactants 87
6.1.6	Alcohol Ethoxylates 87
6.1.7	Alkyl Phenol Ethoxylates 88
6.1.8	Fatty Acid Ethoxylates 88
6.1.9	Sorbitan Esters and Their Ethoxylated Derivatives (Spans and Tweens) 89
6.1.10	Ethoxylated Fats and Oils 90
6.1.11	Amine Ethoxylates 90
6.1.12	Polymeric Surfactants 90
6.1.13	Polyelectrolytes 93
6.1.14	Adsorption of Surfactants at the Solid–Liquid Interface 93
6.1.15	Adsorption of Ionic Surfactants on Hydrophobic Surfaces 94
6.1.16	Adsorption of Ionic Surfactants on Polar Surfaces 97
6.1.17	Adsorption of Nonionic Surfactants 98
6.1.18	Theoretical Treatment of Surfactant Adsorption 101
6.1.19	Examples of Typical Adsorption Isotherms of Model Nonionic Surfactants on Hydrophobic Solids 103
	References 105

7	**Adsorption and Conformation of Polymeric Surfactants at the Solid–Liquid Interface** 107
7.1	Theories of Polymer Adsorption 110
7.2	Experimental Techniques for Studying Polymeric Surfactant Adsorption 117
7.3	Measurement of the Adsorption Isotherm 118
7.4	Measurement of the Fraction of Segments p 118
7.5	Determination of the Segment Density Distribution $\rho(z)$ and Adsorbed Layer Thickness δ_h 119

7.6	Examples of the Adsorption Isotherms of Nonionic Polymeric Surfactants *122*
7.7	Adsorbed Layer Thickness Results *126*
7.8	Kinetics of Polymer Adsorption *128*
	References *129*

8	**Stabilization and Destabilization of Suspensions Using Polymeric Surfactants and the Theory of Steric Stabilization** *131*
8.1	Introduction *131*
8.2	Interaction between Particles Containing Adsorbed Polymeric Surfactant Layers (Steric Stabilization) *131*
8.2.1	Mixing Interaction G_{mix} *132*
8.2.2	Elastic Interaction G_{el} *134*
8.2.3	Total Energy of Interaction *135*
8.2.4	Criteria for Effective Steric Stabilization *135*
8.3	Flocculation of Sterically Stabilized Dispersions *136*
8.3.1	Weak Flocculation *136*
8.3.2	Incipient Flocculation *137*
8.3.3	Depletion Flocculation *138*
8.4	Bridging Flocculation by Polymers and Polyelectrolytes *138*
8.5	Examples for Suspension Stabilization Using Polymeric Surfactants *142*
8.6	Polymeric Surfactants for Stabilization of Preformed Latex Dispersions *146*
	References *148*

9	**Properties of Concentrated Suspensions** *151*
9.1	Interparticle Interactions and Their Combination *151*
9.1.1	Hard-Sphere Interaction *151*
9.1.2	"Soft" or Electrostatic Interaction: Figure 9.1b *152*
9.1.3	Steric Interaction: Figure 9.1c *153*
9.1.4	van der Waals Attraction: Figure 9.1d *156*
9.1.5	Combination of Interaction Forces *157*
9.2	Definition of "Dilute," "Concentrated," and "Solid" Suspensions *160*
9.3	States of Suspension on Standing *164*
	References *169*

10	**Sedimentation of Suspensions and Prevention of Formation of Dilatant Sediments** *171*
10.1	Sedimentation Rate of Suspensions *172*
10.2	Prevention of Sedimentation and Formation of Dilatant Sediments *178*
10.2.1	Balance of the Density of the Disperse Phase and Medium *178*

10.2.2	Reduction of the Particle Size	178
10.2.3	Use of High Molecular Weight Thickeners	178
10.2.4	Use of "Inert" Fine Particles	179
10.2.5	Use of Mixtures of Polymers and Finely Divided Particulate Solids	182
10.2.6	Controlled Flocculation ("Self-Structured" Systems)	183
10.2.7	Depletion Flocculation	186
10.2.8	Use of Liquid Crystalline Phases	190
	References	192

11 Characterization of Suspensions and Assessment of Their Stability 193

11.1	Introduction	193
11.2	Assessment of the Structure of the Solid/Liquid Interface	194
11.2.1	Double-Layer Investigation	194
11.2.1.1	Analytical Determination of Surface Charge	194
11.2.1.2	Electrokinetic and Zeta Potential Measurements	195
11.2.2	Measurement of Surfactant and Polymer Adsorption	196
11.3	Assessment of Sedimentation of Suspensions	199
11.4	Assessment of Flocculation and Ostwald Ripening (Crystal Growth)	201
11.4.1	Optical Microscopy	201
11.4.1.1	Sample Preparation for Optical Microscopy	203
11.4.1.2	Particle Size Measurements Using Optical Microscopy	203
11.4.2	Electron Microscopy	204
11.4.2.1	Transmission Electron Microscopy (TEM)	204
11.4.2.2	Scanning Electron Microscopy (SEM)	204
11.4.3	Confocal Laser Scanning Microscopy (CLSM)	205
11.4.4	Scanning Probe Microscopy (SPM)	205
11.4.5	Scanning Tunneling Microscopy (STM)	206
11.4.6	Atomic Force Microscopy (AFM)	206
11.5	Scattering Techniques	206
11.5.1	Light Scattering Techniques	207
11.5.1.1	Time-Average Light Scattering	207
11.5.2	Turbidity Measurements	208
11.5.3	Light Diffraction Techniques	208
11.5.4	Dynamic Light Scattering – Photon Correlation Spectroscopy (PCS)	211
11.5.5	Backscattering Techniques	214
11.6	Measurement of Rate of Flocculation	214
11.7	Measurement of Incipient Flocculation	215
11.8	Measurement of Crystal Growth (Ostwald Ripening)	216
11.9	Bulk Properties of Suspensions: Equilibrium Sediment Volume (or Height) and Redispersion	216
	References	217

12	**Rheological Techniques for Assessment of Stability of Suspensions** *219*
12.1	Introduction *219*
12.1.1	Steady-State Shear Stress σ–Shear Rate γ Measurements *219*
12.1.2	Constant Stress (Creep) Measurements *219*
12.1.3	Dynamic (Oscillatory) Measurements *220*
12.2	Steady-State Measurements *220*
12.2.1	Rheological Models for Analysis of Flow Curves *220*
12.2.1.1	Newtonian Systems *220*
12.2.1.2	Bingham Plastic Systems *221*
12.2.1.3	Pseudoplastic (Shear Thinning) System *221*
12.2.1.4	Dilatant (Shear Thickening) System *222*
12.2.1.5	Herschel–Bulkley General Model *222*
12.2.2	The Casson Model *222*
12.2.3	The Cross Equation *222*
12.2.4	Time Effects during Flow Thixotropy and Negative (or anti-) Thixotropy *223*
12.3	Constant Stress (Creep) Measurements *225*
12.3.1	Analysis of Creep Curves *226*
12.3.1.1	Viscous Fluid *226*
12.3.1.2	Elastic Solid *226*
12.3.2	Viscoelastic Response *226*
12.3.2.1	Viscoelastic Liquid *226*
12.3.2.2	Viscoelastic Solid *227*
12.3.3	Creep Procedure *228*
12.4	Dynamic (Oscillatory) Measurements *229*
12.4.1	Analysis of Oscillatory Response for a Viscoelastic System *229*
12.4.2	Vector Analysis of the Complex Modulus *230*
12.4.3	Dynamic Viscosity η' *230*
12.4.4	Note that $\eta \to \eta(0)$ as $\omega \to 0$ *230*
12.4.5	Strain Sweep *231*
12.4.6	Oscillatory Sweep *232*
12.4.7	The Cohesive Energy Density E_c *232*
12.4.8	Application of Rheological Techniques for the Assessment and Prediction of the Physical Stability of Suspensions *233*
12.4.8.1	Rheological Techniques for Prediction of Sedimentation and Syneresis *233*
12.4.8.2	Role of Thickeners *235*
12.4.9	Assessment and Prediction of Flocculation Using Rheological Techniques *235*
12.4.9.1	Strain Sweep Measurements *238*
12.4.9.2	Oscillatory Sweep Measurements *239*
12.4.10	Examples of Application of Rheology for Assessment and Prediction of Flocculation *240*

12.4.10.1	Flocculation and Restabilization of Clays Using Cationic Surfactants *240*	
12.4.10.2	Flocculation of Sterically Stabilized Dispersions *240*	
	References *241*	

13 Rheology of Concentrated Suspensions *243*
13.1 Introduction *243*
13.1.1 The Einstein Equation *244*
13.1.2 The Batchelor Equation *244*
13.1.3 Rheology of Concentrated Suspensions *244*
13.1.3.1 Rheology of Hard-Sphere Suspensions *245*
13.1.3.2 Rheology of Systems with "Soft" or Electrostatic Interaction *246*
13.1.3.3 Rheology of Sterically Stabilized Dispersions *248*
13.1.3.4 Rheology of Flocculated Suspensions *250*
13.1.4 Analysis of the Flow Curve *258*
13.1.4.1 Impulse Theory: Goodeve and Gillespie *258*
13.1.4.2 Elastic Floc Model: Hunter and Coworkers *259*
13.1.5 Fractal Concept for Flocculation *259*
13.1.6 Examples of Strongly Flocculated (Coagulated) Suspension *261*
13.1.6.1 Coagulation of Electrostatically Stabilized Suspensions by Addition of Electrolyte *261*
13.1.7 Strongly Flocculated Sterically Stabilized Systems *263*
13.1.7.1 Influence of the Addition of Electrolyte *263*
13.1.7.2 Influence of Increase of Temperature *266*
13.1.8 Models for Interpretation of Rheological Results *267*
13.1.8.1 Dublet Floc Structure Model *267*
13.1.8.2 Elastic Floc Model *268*
  References *270*

Index *271*

"Dedicated to our grandchildren:
Nadia, Dominic, Theodore, Bruno, Viola, Gabriel, Raphael"

Preface

The dispersion of powders, both hydrophobic (such as pharmaceuticals, agrochemicals, organic pigments, and ceramics) and hydrophilic (such as oxides and clays), into liquids, both aqueous and nonaqueous, presents a challenge to most industries. It is essential to understand the process of dispersion at a fundamental level in order to be able to prepare suspensions that are suitable for applications and with a desirable shelf life. Dry powders usually consist of aggregates and agglomerates that need to be dispersed in the liquid to produce "individual" units that may be further subdivided (by comminution or wet milling) into smaller units. This requires understanding of the various phenomena such as powder wetting, dispersion of aggregates, and agglomerates and comminution of the primary particles into smaller units. Once a powder is dispersed into the liquid and the primary particles are broken into smaller units, it is essential to prevent aggregation of the particles, a phenomena described as flocculation or coagulation. This requires the presence of an effective repulsive energy that must overcome the attractive van der Waals energy. It is also essential to prevent particle settling and formation of hard sediments that are very difficult to redisperse. This requires the addition of a "thickener" (rheology modifier) that forms a "gel network" in the continuous phase. The objectives of the present book are to address these phenomena at a fundamental level and to describe the various methods that can be applied to prevent the instability of the suspension. The book starts with a general introduction (Chapter 1) that describes the industrial applications of suspensions. The fundamental knowledge required for successful dispersion of powder into liquid and subsequent stabilization of the suspension is briefly described to give an outline of the book contents. Chapter 2 describes in detail the process of powder wetting with particular emphasis on the contact angle concept, adhesion tension, and spreading coefficient. Chapter 3 describes the critical surface tension of wetting that can be applied to select wetting agents. The role of surfactants in dispersion wetting is described at a fundamental level. The methods of measuring the contact on substrates as well as powders are described. This is followed by a section on the methods that can be applied for assessment of powder wetting. The surfactants that can be used to achieve the most efficient wetting of hydrophobic powders in aqueous media are briefly described. Chapter 4 describes the structure of the solid/liquid interface and electrostatic stabilization. The accepted picture of the electrical double layer is described. This is followed by a description of the repulsive energy that arises from interaction of the electrical double layers on close approach of the particles. The universal van der Waals attraction is analyzed in terms of the London dispersion attraction using the macroscopic Hamaker analysis. The combination of electrostatic repulsion with van der Waal attraction forms the well-known theory of stability of hydrophobic colloids due to Deryaguin–Landau–Verwey–Overbeek (DLVO theory). This explains the kinetic stability of suspensions resulting from the presence of an energy barrier at intermediate separation distance between the particles. The height of this barrier determines the long-term stability of the suspension. It increases with the increase of surface (or zeta) potential, decrease of electrolyte concentration, and valency of the ions. The kinetic process of fast (in the absence of an energy barrier) and slow (in the presence of an energy barrier) flocculation is described and the conditions of stability/instability of suspensions can be well described in terms of a quantitative stability ration. Chapter 5 describes the various electrokinetic phenomena and the concept of zeta potential that is important for describing the stability of the suspension. The various techniques that can be applied for measuring the zeta potential are described. Chapter 6 gives a general classification of dispersing agents and the adsorption of surfactants at the solid/liquid interface. The adsorption of ionic surfactants on hydrophobic and hydrophilic (charged) surfaces is described with particular emphasis on the driving force for adsorption and the structure of the adsorbed layer. This is followed by description of the adsorption of nonionic surfactants on both hydrophobic and hydrophilic surfaces. Chapter 7 describes the adsorption of polymers (polymeric surfactants) at the solid/liquid interface. Particular attention is given to the conformation of the polymer on solid surfaces that is determined by the structure of the polymer. The various techniques that can be applied for measurement of the adsorption parameters, namely the amount of polymer adsorbed, the fraction of segments in direct contact with the surface, and the adsorbed layer thickness, are described. Chapter 8 describes the stabilization of suspensions using polymeric surfactants and steric stabilization. The latter is described in terms of the unfavorable mixing of the stabilizing chains (when these are in good solvent conditions) and the reduction of configurational entropy on overlap of the steric layers. The criteria for effective steric stabilization are described. This is followed by a description of the conditions responsible for weak and strong flocculation. Chapter 9 gives a brief description of concentrated suspension. A distinction could be made between "dilute," "concentrated," and "solid" suspensions in terms of the balance between the Brownian diffusion and interparticle interaction. The various states that are reached on standing could be described and analyzed in term of the interparticle interactions. Chapter 10 deals with the problem of particle sedimentation and formation of "dilatant" sediments. The rate of sedimentation of "dilute," "moderately concentrated," and "concentrated" suspensions is analyzed using the available theories. The methods that can be applied to reduce or eliminate sedimentation are described with particular attention to the use of high molecular weight polymers ("thickeners"). The importance of measuring the suspension viscosity at low shear stresses or shear rates is emphasized to show the correlation of the zero-shear (residual) viscosity with particle sedimentation. Chapter 11 describes the various techniques that can be applied to characterize the suspension after its dilution with the continuous medium. Both optical and electron microscopy can be used for quantitative measurement of the article size distribution. Specialized techniques such as confocal laser microscopy and atomic force microscopy could also be applied. The various light scattering methods that can be applied for characterization of the suspension are also described. Chapter 12 describes the methods that can be applied for characterization of the suspension without dilution. The most convenient methods are based on rheological techniques, namely steady state, constant stress (creep), and oscillatory. A description of the principles of each method is given. This is followed by application of these rheological methods for the assessment and prediction of the long-term physical stability of the suspension. Chapter 13 gives an account of the rheology of concentrated suspensions. The main factors that affect the rheology of a suspension, namely the Brownian diffusion, hydrodynamic and interparticle interaction, are described. A summary is given on the rheology of hard sphere, electrostatically stabilized, sterically stabilized, and flocculated suspensions.

In summary this book gives a comprehensive account on the fundamental principles involved in dispersion of powders into liquids as well as stabilization of the resulting suspension. It should serve as a valuable text to those scientists in the industry who deal regularly with the formulation of suspensions. It can also be valuable to graduate and postgraduate workers in academia who are carrying out research on the subject of suspensions.

November 2011 *Tharwat F. Tadros*

1
General Introduction

The dispersion of powders into liquids is a process that occurs in many industries of which we mention paints, dyestuffs, paper coatings, printing inks, agrochemicals, pharmaceuticals, cosmetics, food products, detergents, and ceramics. The powder can be hydrophobic such as organic pigments, agrochemicals, and ceramics or hydrophilic such as silica, titania, and clays. The liquid can be aqueous or nonaqueous.

The dispersion of a powder in a liquid is a process whereby aggregates and agglomerates of powders are dispersed into "individual" units, usually followed by a wet milling process (to subdivide the particles into smaller units) and stabilization of the resulting dispersion against aggregation and sedimentation [1–3]. This is illustrated in Figure 1.1.

The powder is considered hydrophobic if there is no affinity between its surface and water, for example, carbon black, many organic pigments, and some ceramic powders such as silicon carbide or silicon nitride. In contrast a hydrophilic solid has strong affinity between its surface and water, for example, silica, alumina, and sodium montmorillonite clay.

1.1
Fundamental Knowledge Required for Successful Dispersion of Powders into Liquids

Several fundamental processes must be considered for the dispersion process and these are summarized below.

1.1.1
Wetting of Powder into Liquid

This is determined by surface forces whereby the solid/air interface characterized by an interfacial tension (surface energy) γ_{SA} is replaced by the solid/liquid interface characterized by an interfacial tension (surface energy) γ_{SL} [1]. Polar (hydrophilic) surfaces such as silica or alumina have high surface energy and hence they can be easily wetted in a polar liquid such as water. In contrast nonpolar

Dispersion of Powders in Liquids and Stabilization of Suspensions, First Edition. Tharwat F. Tadros.
© 2012 Wiley-VCH Verlag GmbH & Co. KGaA. Published 2012 by Wiley-VCH Verlag GmbH & Co. KGaA.

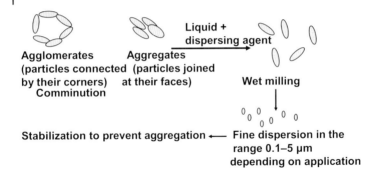

Figure 1.1 Schematic representation of the dispersion process.

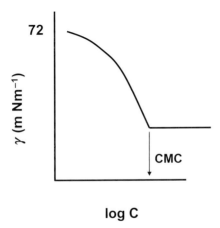

Figure 1.2 Surface tension–log C curve.

(hydrophobic) surfaces such as carbon black and many organic pigments have low surface energy and hence they require a surface active agent (surfactant) in the aqueous phase to aid wetting. The surfactant lowers the surface tension γ of water from ~72 to ~ 30–40 mN m^{-1} depending on surfactant nature and concentration. This is illustrated in Figure 1.2, which shows the γ–log C (where C is the surfactant concentration) relationship of surfactant solutions. It can be seen that γ decreases gradually with the increase in surfactant concentration, and above a certain concentration it shows a linear decrease with the increase in log C. Above a critical surfactant concentration γ remains constant. This critical concentration is that above which any added surfactant molecules aggregate to form micelles that are in equilibrium with the surfactant monomers. This critical concentration is referred to as the critical micelle concentration (CMC).

There are generally two approaches for treating surfactant adsorption at the A/L interface. The first approach, adopted by Gibbs, treats adsorption as an equilibrium phenomenon whereby the second law of thermodynamics may be applied using surface quantities. The second approach, referred to as the equation-of-state approach, treats the surfactant film as a two-dimensional layer with a surface pressure π that may be related to the surface excess Γ (amount of surfactant adsorbed per unit area). Below, the Gibbs treatment that is commonly used to describe adsorption at the A/L interface is summarized.

Gibbs [4] derived a thermodynamic relationship between the surface or interfacial tension γ and the surface excess Γ (adsorption per unit area). The starting point of this equation is the Gibbs–Deuhem equation. At constant temperature, and in the presence of adsorption, the Gibbs–Deuhem equation is

$$d\gamma = -\sum \frac{n_i^\sigma}{A} d\mu_i = -\sum \Gamma_i \, d\mu_i \tag{1.1}$$

where $\Gamma_i = n_i^\sigma / A$ is the number of moles of component i and adsorbed per unit area.

Equation (1.1) is the general form for the Gibbs adsorption isotherm. The simplest case of this isotherm is a system of two components in which the solute (2) is the surface active component, that is, it is adsorbed at the surface of the solvent (1). For such a case, Eq. (1.1) may be written as

$$-d\gamma = \Gamma_1^\sigma \, d\mu_1 + \Gamma_2^\sigma \, d\mu_2 \tag{1.2}$$

and if the Gibbs dividing surface is used, $\Gamma_1 = 0$ and,

$$-d\gamma = \Gamma_{1,2}^\sigma \, d\mu_2 \tag{1.3}$$

where $\Gamma_{2,1}^\sigma$ is the relative adsorption of (2) with respect to (1). Since

$$\mu_2 = \mu_2^o + RT \ln a_2^l \tag{1.4}$$

or

$$d\mu_2 = RT \, d \ln a_2^l \tag{1.5}$$

then

$$-d\gamma = \Gamma_{2,1}^\sigma RT \, d \ln a_2^l \tag{1.6}$$

or

$$\Gamma_{2,1}^\sigma = -\frac{1}{RT} \left(\frac{d\gamma}{d \ln a_2^l} \right) \tag{1.7}$$

where a_2^l is the activity of the surfactant in bulk solution that is equal to $C_2 f_2$ or $x_2 f_2$, where C_2 is the concentration of the surfactant in mol dm^{-3} and x_2 is its mole fraction.

Equation (1.7) allows one to obtain the surface excess (abbreviated as Γ_2) from the variation of surface or interfacial tension with surfactant concentration. Note that $a_2 \sim C_2$ since in dilute solutions $f_2 \sim 1$. This approximation is valid since most surfactants have low c.m.c. (usually less than 10^{-3} mol dm^{-3}) but adsorption is complete at or just below the c.m.c.

The surface excess Γ_2 can be calculated from the linear portion of the γ–log C_2 curves before the c.m.c. Such a γ–log C curve is illustrated in Figure 1.2 for the air/water interface. As mentioned above, Γ_2 can be calculated from the slope of the linear position of the curves shown in Figure 1.2 just before the c.m.c. is reached. From Γ_2, the area per surfactant ion or molecule can be calculated since

$$\text{Area/molecule} = \frac{1}{\Gamma_2 N_{av}} \tag{1.8}$$

where N_{av} is Avogadro's constant. Determining the area per surfactant molecule is very useful since it gives information on surfactant orientation at the interface. For example, for ionic surfactants such as sodium dodecyl sulfate, the area per surfactant is determined by the area occupied by the alkyl chain and head group if these molecules lie flat at the interface, whereas for vertical orientation, the area per surfactant ion is determined by that occupied by the charged head group, which at low electrolyte concentration will be in the region of $0.40\,nm^2$. Such an area is larger than the geometrical area occupied by a sulfate group, as a result of the lateral repulsion between the head group. On addition of electrolytes, this lateral repulsion is reduced and the area/surfactant ion for vertical orientation will be lower than $0.4\,nm^2$ (reaching in some case $0.2\,nm^2$). On the other hand, if the molecules lie flat at the interface, the area per surfactant ion will be considerably higher than $0.4\,nm^2$.

Another important point can be made from the γ–log C curves. At concentration just before the break point, one has the condition of constant slope, which indicates that saturation adsorption has been reached. Just above the break point,

$$\left(\frac{\partial \gamma}{\partial \ln a_2}\right)_{p,T} = \text{constant} \tag{1.9}$$

$$\left(\frac{\partial \gamma}{\partial \ln a_2}\right)_{p,T} = 0 \tag{1.10}$$

indicating the constancy of γ with log C above the c.m.c. Integration of Eq. (1.10) gives

$$\gamma = \text{constant} \times \ln a_2 \tag{1.11}$$

Since γ is constant in this region, then a_2 must remain constant. This means that the addition of surfactant molecules, above the c.m.c., must result in association with form units (micellar) with low activity.

The hydrophilic head group may be unionized, for example, alcohols or poly(ethylene oxide) alkane or alkyl phenol compounds, weakly ionized such as carboxylic acids, or strongly ionized such as sulfates, sulfonates, and quaternary ammonium salts. The adsorption of these different surfactants at the air/water interface depends on the nature of the head group. With nonionic surfactants, repulsion between the head groups is small and these surfactants are usually strongly adsorbed at the surface of water from very dilute solutions. Nonionic surfactants have much lower c.m.c. values when compared with ionic surfactants with the same alkyl chain length. Typically, the c.m.c. is in the region of 10^{-5}–10^{-4} mol dm^{-3}. Such nonionic surfactants form closely packed adsorbed layers at concentrations lower than their c.m.c. values. The activity coefficient of such surfactants is close to unity and is only slightly affected by the addition of moderate amounts of electrolytes (or change in the pH of the solution). Thus, nonionic surfactant adsorption is the simplest case since the solutions can be represented by a two-component system and the adsorption can be accurately calculated using Eq. (1.7).

With ionic surfactants, on the other hand, the adsorption process is relatively more complicated since one has to consider the repulsion between the head groups and the effect of the presence of any indifferent electrolyte. Moreover, the Gibbs adsorption equation has to be solved taking into account the surfactant ions, the counterion, and any indifferent electrolyte ions present. For a strong surfactant electrolyte such as Na$^+$R$^-$,

$$\Gamma_2 = \frac{1}{2RT}\frac{d\gamma}{d\ln a_\pm} \tag{1.12}$$

The factor of 2 in Eq. (1.12) arises because both surfactant ion and counterion must be adsorbed to maintain neutrally, and $d\gamma/d\ln a_\pm$ is twice as large as for an un-ionized surfactant.

If a nonadsorbed electrolyte, such as NaCl, is present in large excess, then any increase in the concentration of Na$^+$R$^-$ produces a negligible increase in the Na$^+$ ion concentration and therefore $d\mu_{Na}$ becomes negligible. Moreover, $d\mu_{Cl}$ is also negligible, so the Gibbs adsorption equation reduces to

$$\Gamma_2 = -\frac{1}{RT}\left(\frac{\partial \gamma}{\partial \ln C_{NaR}}\right) \tag{1.13}$$

that is, it becomes identical to that for a nonionic surfactant.

The above discussion clearly illustrates that for the calculation of Γ_2 from the γ–log C curve one has to consider the nature of the surfactant and the composition of the medium. For nonionic surfactants the Gibbs adsorption (Eq. (1.7)) can be directly used. For ionic surfactant, in the absence of electrolytes the right-hand side of Eq. (1.7) should be divided by 2 to account for surfactant dissociation. This factor disappears in the presence of the high concentration of an indifferent electrolyte.

Surfactants also adsorb on hydrophobic surfaces with the hydrophobic group pointing to the surface and the hydrophilic group pointing to water. Adsorption increases with the increase in surfactant concentration reaching a limiting value (the saturation adsorption) near the critical micelle concentration. This is illustrated in Figure 1.3. With many solid/surfactant systems the adsorption follows the Langmuir theory.

The adsorption of ionic surfactants on hydrophobic surfaces may be represented by the Stern–Langmuir isotherm [5]. Consider a substrate containing N_s sites (mol m^{-2}) on which Γ mol m^{-2} of surfactant ions are adsorbed. The surface coverage θ is (Γ/N_s) and the fraction of uncovered surface is $(1 - \theta)$.

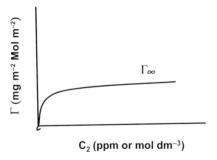

Figure 1.3 Langmuir-type adsorption isotherm.

The rate of adsorption is proportional to the surfactant concentration expressed in mole fraction, $(C/55.5)$, and the fraction of free surface $(1 - \theta)$, that is,

$$\text{Rate of adsorption} = k_{ads}\left(\frac{C}{55.5}\right)(1-\theta) \tag{1.14}$$

where k_{ads} is the rate constant for adsorption.

The rate of desorption is proportional to the fraction of surface covered θ,

$$\text{Rate of desorption} = k_{des}\theta \tag{1.15}$$

At equilibrium, the rate of adsorption is equal to the rate of desorption and the k_{ads}/k_{des} ratio is the equilibrium constant K, that is,

$$\frac{\theta}{(1-\theta)} = \frac{C}{55.5}K \tag{1.16}$$

The equilibrium constant K is related to the standard free energy of adsorption by

$$-\Delta G^\circ_{ads} = RT \ln K \tag{1.17}$$

R is the gas constant and T is the absolute temperature. Equation (1.17) can be written in the form

$$K = \exp\left(-\frac{\Delta G^\circ_{ads}}{RT}\right) \tag{1.18}$$

Combining Eqs. (1.3) and (1.5),

$$\frac{\theta}{1-\theta} = \frac{C}{55.5}\exp\left(-\frac{\Delta G^\circ_{ads}}{RT}\right) \tag{1.19}$$

Equation (1.6) applies only at low surface coverage ($\theta < 0.1$) where lateral interaction between the surfactant ions can be neglected.

At high surface coverage ($\theta > 0.1$) one should take the lateral interaction between the chains into account, by introducing a constant A, for example, using the Frumkin–Fowler–Guggenheim equation [5],

$$\frac{\theta}{(1-\theta)}\exp(A\theta) = \frac{C}{55.5}\exp\left(-\frac{\Delta G^0_{ads}}{RT}\right) \qquad (1.20)$$

Various authors [6, 7] have used the Stern–Langmuir equation in a simple form to describe the adsorption of surfactant ions on mineral surfaces,

$$\Gamma = 2rC\exp\left(-\frac{\Delta G^0_{ads}}{RT}\right) \qquad (1.21)$$

Various contributions to the adsorption free energy may be envisaged. To a first approximation, these contributions may be considered to be additive. In the first instance, ΔG_{ads} may be taken to consist of two main contributions, that is,

$$\Delta G_{ads} = \Delta G_{elec} + \Delta G_{spec} \qquad (1.22)$$

where ΔG_{elec} accounts for any electrical interactions and ΔG_{spec} is a specific adsorption term which contains all contributions to the adsorption free energy that are dependent on the "specific" (nonelectrical) nature of the system [5]. Several authors subdivided ΔG_{spec} into supposedly separate independent interactions [6, 7], for example,

$$\Delta G_{spec} = \Delta G_{cc} + \Delta G_{cs} + \Delta G_{hs} + \cdots \qquad (1.23)$$

where ΔG_{cc} is a term that accounts for the cohesive chain–chain interaction between the hydrophobic moieties of the adsorbed ions, ΔG_{cs} is the term for the chain/substrate interaction whereas ΔG_{hs} is a term for the head group/substrate interaction. Several other contributions to ΔG_{spec} may be envisaged, for example, ion–dipole, ion–induced dipole, or dipole–induced dipole interactions.

Since there is no rigorous theory that can predict adsorption isotherms, the most suitable method to investigate adsorption of surfactants is to determine the adsorption isotherm. Measurement of surfactant adsorption is fairly straightforward. A known mass m (g) of the particles (substrate) with known specific surface area A_s (m^2 g^{-1}) is equilibrated at constant temperature with surfactant solution with initial concentration C_1. The suspension is kept stirred for sufficient time to reach equilibrium. The particles are then removed from the suspension by centrifugation and the equilibrium concentration C_2 is determined using a suitable analytical method. The amount of adsorption Γ (mol m^{-2}) is calculated as follows:

$$\Gamma = \frac{(C_1 - C_2)}{mA_s} \qquad (1.24)$$

The adsorption isotherm is represented by plotting Γ versus C_2. A range of surfactant concentrations should be used to cover the whole adsorption process, that is, from the initial low values to the plateau values. To obtain accurate results, the solid should have a high surface area (usually > 1 m^2).

It is essential to wet both the external and internal surfaces (pores inside agglomerates).

Wetting of the external surface requires surfactants that lower the liquid/air interfacial tension, γ_{LA}, efficiently, in particular under dynamic conditions (dynamic surface tension measurements are more informative). Wetting of the internal surface requires penetration of the liquid into the pores that is determined by the capillary pressure which is directly proportional to γ_{LA}.

A useful concept for assessment of powder wetting is to measure the contact angle θ at the solid/liquid interface, which when combined with the surface tension γ_{LA} can give a quantitative measure of wetting and penetration of the liquid into pores.

1.1.2
Breaking of Aggregates and Agglomerates into Individual Units

This usually requires the application of mechanical energy. High-speed mixers (which produce turbulent flow) are efficient in breaking up the aggregates and agglomerates, for example, Silverson mixers, UltraTurrax. The mixing conditions have to be optimized: heat generation at high stirring speeds must be avoided. This is particularly the case when the viscosity of the resulting dispersion increases during dispersion (note that the energy dissipation as heat is given by the product of the square of the shear rate and the viscosity of the suspension). One should avoid foam formation during dispersion; proper choice of the dispersing agent is essential and antifoams (silicones) may be applied during the dispersion process.

In order to maintain the particles as individual units, it is essential to use a dispersing agent that must provide an effective repulsive barrier preventing aggregation of the particles by van der Waals forces. This dispersing agent must be strongly adsorbed on the particle surface and should not be displaced by the wetting agent. The repulsive barrier can be electrostatic in nature, whereby electrical double layers are formed at the solid/liquid interface [8, 9]. These double layers must be extended (by maintaining low electrolyte concentration) and strong repulsion occurs on double-layer overlap. Alternatively, the repulsion can be produced by the use of nonionic surfactant or polymer layers which remain strongly hydrated (or solvated) by the molecules of the continuous medium [10]. On approach of the particles to a surface-to-surface separation distance that is lower than twice the adsorbed layer thickness, strong repulsion occurs as a result of two main effects: (i) unfavorable mixing of the layers when these are in good solvent conditions and (ii) loss of configurational entropy on significant overlap of the adsorbed layers. This process is referred to as steric repulsion. A third repulsive mechanism is that whereby both electrostatic and steric repulsion are combined, for example, when using polyelectrolyte dispersants.

1.1.3
Wet Milling or Comminution

The primary particles produced after dispersion are subdivided into smaller units by milling or comminution (a process that requires rupture of bonds). Wet milling can be achieved using ball mills, bead mills (ceramic balls or beads are normally used to avoid contamination), or colloid mills. Again the milling conditions must be adjusted to prevent heat and/or foam formation. The role of the dispersing agent (surfactant) in breaking the primary particles is usually described in terms of the "Rehbinder" effect, that is, adsorption of the dispersing agent molecules on the surface of the particles (which lowers their surface energy) and in particular in the "cracks" which facilitate their propagation.

1.1.4
Stabilization of the Resulting Dispersion

The particles of the resulting dispersion may undergo aggregation (flocculation) on standing as a result of the universal van der Waals attraction. Any two macroscopic bodies (such as particles) in a dispersion attract each other as a result of the London dispersion attractive energy between the particles. This attractive energy becomes very large at short distances of separation between the particles. As mentioned above, to overcome the everlasting van der Waals attraction energy, it is essential to have a repulsive energy between the particles. Two main repulsive energies can be described: electrostatic repulsive energy is produced by the presence of electrical double layers around the particles produced by charge separation at the solid/liquid interface. The dispersant should be strongly adsorbed to the particles, produce high charge (high surface or zeta potential), and form an extended double layer (that can be achieved at low electrolyte concentration and low valency) [8, 9]. The second repulsive energy, steric repulsive energy, is produced by the presence of adsorbed (or grafted) layers of surfactant or polymer molecules. In this case the nonionic surfactant or polymer (referred to as polymeric surfactant) should be strongly adsorbed to the particle surface and the stabilizing chain should be strongly solvated (hydrated in the case of aqueous suspensions) by the molecules of the medium [10]. The most effective polymeric surfactants are those of the A–B, A–B–A block, or BA_n graft copolymer. The "anchor" chain B is chosen to be highly insoluble in the medium and has strong affinity to the surface. The A stabilizing chain is chosen to be highly soluble in the medium and strongly solvated by the molecules of the medium. For suspensions of hydrophobic solids in aqueous media, the B chain can be polystyrene, poly(methylmethacrylate), or poly(propylene oxide). The A chain could be poly(ethylene oxide) which is strongly hydrated by the medium.

1.1.5
Prevention of Ostwald Ripening (Crystal Growth)

The driving force for Ostwald ripening is the difference in solubility between the small and large particles (the smaller particles have higher solubility than the larger ones). The difference in chemical potential between different sized particles was given by Lord Kelvin [11],

$$S(r) = S(\infty)\exp\left(\frac{2\sigma V_m}{rRT}\right) \tag{1.25}$$

where $S(r)$ is the solubility surrounding a particle of radius r, $S(\infty)$ is the bulk solubility, σ is the solid/liquid interfacial tension, V_m is the molar volume of the dispersed phase, R is the gas constant, and T is the absolute temperature. The quantity $(2\sigma V_m/rRT)$ is termed the characteristic length. It has an order of ~1 nm or less, indicating that the difference in solubility of a 1 μm particle is of the order of 0.1% or less. Theoretically, Ostwald ripening should lead to condensation of all particles into a single. This does not occur in practice since the rate of growth decreases with the increase of the particle size.

For two particles with radii r_1 and r_2 ($r_1 < r_2$),

$$\frac{RT}{V_m}\ln\left[\frac{S(r_1)}{S(r_2)}\right] = 2\sigma\left[\frac{1}{r_1} - \frac{1}{r_2}\right] \quad (1.26)$$

Equation (1.26) shows that the larger the difference between r_1 and r_2, the higher the rate of Ostwald ripening.

Ostwald ripening can be quantitatively assessed from plots of the cube of the radius versus time t,

$$r^3 = \frac{8}{9}\left[\frac{S(\infty)\sigma V_m D}{\rho RT}\right]t \quad (1.27)$$

where D is the diffusion coefficient of the disperse phase in the continuous phase.

Several factors affect the rate of Ostwald ripening and these are determined by surface phenomena, although the presence of surfactant micelles in the continuous phase can also play a major role. Trace amounts of impurities that are highly insoluble in the medium and have strong affinity to the surface can significantly reduce Ostwald ripening by blocking the active sites on the surface on which the molecules of the active ingredient can deposit. Many polymeric surfactants, particularly those of the block and graft copolymer types, can also reduce the Ostwald ripening rate by strong adsorption on the surface of the particles, thus making it inaccessible for molecular deposition. Surfactant micelles that can solubilize the molecules of the active ingredient may enhance the rate of crystal growth by increasing the flux of transport by diffusion.

1.1.6
Prevention of Sedimentation and Formation of Compact Sediments (Clays)

Sedimentation is the result of gravity – the particle density is usually larger than that of the medium. The particles tend to remain uniformly dispersed as a result of their Brownian (thermal) motion (of the order of kT; k is the Boltzmann constant and T is the absolute temperature). The gravity force is $(4/3)\pi R^3 \Delta\rho g L$ (R is the particle radius, $\Delta\rho$ is the buoyancy or the density difference between the particle and the medium, g is the acceleration due to gravity, and L is the length of the container). When $(4/3)\pi R^3 \Delta\rho g L > kT$, sedimentation of the individual particles will occur. The particles in the sediment will rotate around each other (as a result of the repulsive forces between them) producing a compact sediment (technically referred to as a "clay"). The compact sediments are very difficult to redisperse (due to the small distances between the particles). Production of "clays" must be prevented by several processes [3]: using "thickeners" which can produce a "gel" network in the continuous phase. At low stresses (which are exerted by the particles) these "gel" networks produce a high viscosity preventing particle sedimentation. In most cases, the particles and the "thickener" produce a "three-dimensional" structure that prevents separation of the dispersion. In some cases, "controlled" flocculation of the particles (self-structured systems) may be used to prevent sedimentation. The most widely used "thickeners" for prevention of sedimentation in aqueous suspensions are high molecular weight water soluble polymers such as hydroxyethylcellulose (HEC) or xanthan gum (a polysaccharide with a molecular weight >10^6). These polymers show non-Newtonian (shear thinning) behavior above a critical concentration C^* at which polymer coil overlap occurs. Above C^* the residual (or zero shear) viscosity show a rapid increase with the further increase in polymer concentration. These overlapped coils form a "three-dimensional" gel network in the continuous phase, thus preventing particle sedimentation. Alternatively, one can use finely divided "inert" particles such as swellable clays (e.g., sodium montmorillonite) or silica that can also produce a three-dimensional gel network in the continuous phase. In most cases, a mixture of high molecular weight polymer such as xanthan gum and sodium montmorillonite is used. This gives a more robust gel structure that is less temperature dependent.

1.2
Particle Dimensions in Suspensions

It is necessary to define the lower and upper limit of particle dimensions. This is by no means exact, and only an arbitrary range of dimensions may be chosen depending on the range of properties of the system with the change of size. For example, the lower limit may be set by the smallest aggregate for which it is meaningful to distinguish between "surface" and "interior" molecules; this is arbitrarily taken to be about 1 nm. Simple considerations show that when matter is subdivided into particles with dimensions below 1000 nm (1 μm) a substantial proportion of the atoms or molecules come close to the surface and make contributions to the energy that differ from those made by the molecules in the interior. This is sometimes set as the upper limit of the colloidal state, and therefore solid/liquid dispersions within the size range 1 nm–1 μm may be referred to as colloidal suspensions. It is difficult to set an upper limit for the size of particles in suspensions, but generally speaking particles of diameters tens of micrometers may be encountered in many practical systems.

1.3
Concentration Range of Suspensions

The particle concentration in a suspension is usually described in terms of its volume fraction ϕ, that is, the total volume of the particles divided by the total volume of the suspensions. Volume fractions covering a wide range (0.01–0.7 or higher) are encountered in many practical systems. It is difficult to define an exact value for ϕ at which a suspension may be considered to be "dilute" or "concentrated." The most convenient way is to consider the balance between the particles' translational motion and their interparticle interactions. At one extreme, a suspension may be considered "dilute" if the thermal motion of the particles predominate over the imposed interparticle forces [12, 13]. In this case the particle translational motion is large, and only occasional contacts occur between the particles, that is, the particles do not "see" each other until collision occurs, giving a random arrangement of particles. In this case the particle interactions can be described by two-body collisions. Such dilute suspensions show no phase separation when the particle sizes are in the colloid range and the density of the particles is not significantly larger than that of the medium (e.g., polystyrene latex suspension). Moreover, the properties of the suspension are time independent, and therefore any time-averaged quantity such as viscosity or light scattering can be extrapolated to infinite dilution to obtain the particle size.

As the particle number is increased in a suspension, the volume of space occupied by the particles increases relative to the total volume of the suspension and a proportion of space is excluded in terms of its occupancy by a single particle. Moreover, the probability of particle–particle interaction increases and the forces of interaction between the particles play a dominant role in determining the properties of the system. With the further increase in particle number concentration, the interactive contact between the particles produces a specific order between them, and a highly developed structure is obtained. Such ordered systems are referred to as "solid" suspensions. In such cases, any particle in the system interacts with many neighbors. The particles are only able to vibrate within a distance that is small relative to the particle radius. The vibrational amplitude is essentially time independent, and hence the properties of the suspension such as its elastic modulus are also time independent.

In between the random arrangement of particles in "dilute" suspensions and the highly ordered structure of "solid" suspensions, one may loosely define concentrated suspensions [12, 13]. In this case, the particle interactions occur by many-body collisions and the translational motion of the particles is restricted. However, the translational motion of the particles is not reduced to the same extent as with "solid" suspensions, that is, the vibrational motion of the particles in this case is large compared with the particle radius. A time-dependent system arises in which there will be spatial and temporal correlations.

1.4
Outline of the Book

The text is organized as follows: Chapter 2 deals with the fundamentals of wetting with particular reference to the contact angle concept. The thermodynamic treatment of the contact angle and Young's equation is presented. This is followed by the analysis of the spreading pressure, adhesion tension, work of adhesion, and cohesion. The Harkins definition of spreading coefficient is discussed. Finally, the contact angle hysteresis and its reasons are discussed in terms of surface roughness and surface heterogeneity.

Chapter 3 deals with the concept of critical surface tension of wetting, its measurement, and its value in characterizing solid surfaces. The role of surfactants on powder wetting is analyzed in terms of its adsorption and effect on the contact angle. A distinction is made between the dynamic and equilibrium processes of surfactant adsorption. A section is given on the analysis of the dynamics of adsorption and its measurement. An analysis is given of the process of dispersion wetting. The process of wetting of the internal surface and the capillary phenomena is described. This is followed by the analysis of the rate of penetration of liquids into pores between aggregates and agglomerates. The assessment of wettability using sinking time test and contact angle measurements is described. This is followed by classification of wetting agents for hydrophobic solids in aqueous media.

Chapter 4 deals with the structure of the solid/liquid interface and electrostatic stabilization. It starts with the description of the origin of charge on surfaces and creation of the electrical double layer. The structure of the electrical double layer following Gouy–Chapman–Stern–Grahame pictures is given. Analysis of the double-layer extension and electrostatic repulsion is described at a fundamental level. The effect of electrolyte concentration and valency on double-layer extension and repulsion is described. Analysis of the van der Waals attraction is given in terms of the London dispersion forces. The expressions for the London van der Waals attraction and the effect of the medium are described. The combination of electrostatic repulsion and van der Waals attraction gives the total energy of interaction between particles as a function of their surface-to-surface separation. This forms the basis of the theory of colloid stability due to Deryaguin–Landau–Verwey–Overbeek (DLVO) theory. Energy–distance curves are given with particular reference to the effect of electrolyte concentration and valency. The main criteria for effective electrostatic stabilization I are given. This is followed by a section on suspension flocculation as a kinetic process. Both fast and slow flocculation are described followed by the concept of stability ratio. This leads to the definition of critical coagulation concentration (CCC) and its dependence on electrolyte valency as described by the Schultze–Hardy rule.

Chapter 5 deals with the electrokinetic phenomena and the zeta potential. The process of charge separation in the region between two adjoining phases is described. The arrangement of charges on one phase and the distribution of charges in the adjacent phase result in the formation of the electrical double layer described in Chapter 4. When one of these phases is caused to move tangentially past the second phase leads to the phenomena of electrokinetic effects which can be classified into four main topics, namely, electrophoresis, electro-osmosis, streaming potential, and sedimentation potential. A brief description of each of these effects is given. Particular emphasis is given to the process of electrophoresis that is commonly applied for suspensions. The concept of surface of shear is described and hence the definition of electrokinetic or zeta potential. The calculation of the zeta potential from electrophoretic mobility using Smoluchowski, Huckel, Henry and Wiersema, Loeb, and Overbeek theories is described. This is followed by the experimental techniques of measurement of electrophoretic mobility and zeta potential. Both microelectrophoresis and electrophoretic light scattering (Laser-Doppler method) or laser velocimetry are described.

Chapter 6 gives the general classification of dispersing agents and adsorption of surfactants at the solid/liquid interface. The dispersing agents can be ionic or nonionic surfactants, polymeric surfactants, and polyelectrolytes. The process of surfactant adsorption is described at a fundamental level using the Stern–Langmuir analysis of surfactant adsorption and its modification by Frumkin–Fowler–Guggenheim for high surface coverage. The adsorption of ionic and nonionic surfactants on hydrophobic and hydrophilic solids is described with emphasis on the free energy of surfactant adsorption. The concept of hemi-micelle formation on solid surfaces is described. This is followed by the different adsorption isotherms of nonionic surfactants on solid surfaces and the structure of the adsorbed layers.

Chapter 7 deals with the process of adsorption of polymers at the solid/liquid interface. The complexity of the process of polymer adsorption and the importance of the configuration (conformation) of the polymer at the solid/liquid interface are described with particular reference to the polymer/surface and polymer/solvent interaction. The conformation of homopolymers, and block and graft copolymers at the solid/liquid interface is described. For characterization of polymer adsorption one needs to determine the various parameters that determine the process. The theories of polymer adsorption are briefly described. This is followed by the description of the experimental methods for determination of the various adsorption parameters.

Chapter 8 deals with the process of stabilization of suspensions using polymeric surfactants and the theory of steric stabilization. The interaction between particles containing adsorbed polymer layers is described in terms of interpenetration and/or compression of the adsorbed layers. The unfavorable mixing of the stabilizing chains when these are in good solvent conditions is described at a fundamental level. The entropic, volume restriction, or elastic interaction is described. The combination of mixing and elastic interaction gives the total steric interaction. The combination of steric repulsion with van der Waals attraction shows the resulting energy–distance curve for sterically stabilized suspensions. The role of the adsorbed layer thickness in determining the energy–distance curve is described. This is followed by the main criteria for effective steric stabilization. The conditions for flocculation of sterically stabilized dispersions are described. Both weak (reversible) and strong (incipient) flocculation can be produced depending on the conditions. Particular attention is given to the role of the solvency of the medium for the stabilizing chains.

Chapter 9 deals with the properties of concentrated suspensions. A distinction can be made between "dilute," "concentrated," and "solid" suspension in terms of the balance between Brownian diffusion and interparticle forces as discussed above. The states of suspensions on standing are described in terms of interparticle interactions and the effect of gravity. Three main systems can be distinguished: colloidally stable, coagulated, and weakly flocculated suspensions. The colloidally stable systems are those where the net interaction between the particles is repulsive, whereas coagulated systems are those with net attraction between the particles. Weakly flocculated suspensions are produced with a relatively smaller attraction and in this case the flocculation is reversible. Three systems could be distinguished: weakly flocculated suspension with net attraction in the secondary (shallow) minimum, flocs produced by bridging with polymer chains that are weakly adsorbed on the particle surfaces, and weakly flocculated suspensions produced by addition of "free" (nonadsorbing) polymer in the continuous phase.

Chapter 10 deals with sedimentation of suspensions and prevention of formation of dilatant sediments. It starts with the effect of particle size and its distribution on sedimentation. The sedimentation of very dilute suspensions with a volume fraction $\phi \leq 0.01$ and application of the Stokes law is described. This is followed by the description of sedimentation of moderately concentrated suspensions (with $0.2 \geq \phi \geq 0.1$) and the effect of hydrodynamic interaction; sedimentation of concentrated suspensions ($\phi > 0.2$) and models for its description; sedimentation in non-Newtonian liquids and correlation of sedimentation rate with residual (zero shear) viscosity; and role of thickeners (rheology modifiers) in prevention of sedimentation: balance of density, reduction of particle size, and use of thickeners and finely divided inert particles. The application of depletion flocculation for reduction of sedimentation is described. The use of liquid crystalline phases for reduction of sedimentation is also described.

Chapter 11 deals with characterization of suspensions and assessment of their stability. It gives a brief description of the various techniques that can be applied for measurement of the particle size distribution. This starts with optical microscopy and improvements using phase contrast, differential interference contrast, and polarizing microscopy. The use of electron and scanning electron microscopy for assessment of suspensions is also described. Techniques such as confocal scanning laser microscopy and atomic force microscopy are also briefly described. The various scattering techniques that can be applied for particle size determination are described. They start with application time average and dynamic light scattering. The use of light diffraction techniques that are commonly in practice is also discussed. For assessment of concentrated suspensions backscattering techniques are described.

Chapter 12 describes the methods of evaluation of suspensions without dilution, in particular the application of rheological techniques. The various rheological methods that can be applied are described. The first method is steady-state shear stress–shear rate measurements that can distinguish between Newtonian and non-Newtonian flow. The various rheological models that can be applied for the analysis of the flow curves are described. This is followed by the description of constant stress (creep) measurements and the concept of residual (zero shear) viscosity and critical stress. The last part deals with dynamic or oscillatory techniques and calculation of the complex, storage, and loss moduli. This allows one to obtain information on the structure of the suspension and the cohesive energy density of the flocculated structure. A section is devoted to the use of rheology for the assessment and prediction of the physical stability of the suspension. Chapter 13 describes the rheology of concentrated suspensions which depends on the balance between Brownian diffusion, hydrodynamic interaction, and interparticle forces. An important dimensionless number is the ratio between the relaxation time of the suspension and the applied experimental time and this is defined as the Deborah number. The Einstein equation for very dilute suspensions ($\phi \leq 0.01$) and its modification by Batchelor for moderately concentrated suspensions (with $0.2 \geq \phi \geq 0.1$) are described. Rheology of concentrated suspensions ($\phi > 0.2$) is described. It starts with hard-sphere suspensions where both repulsion and attraction are screened. The models for the analysis of the viscosity versus volume fraction curves are described. This is followed by electrostatically stabilized suspensions where the rheology is determined by double-layer repulsion. The rheology of sterically stabilized suspensions is described with particular reference to the effect of the adsorbed layer thickness. Finally, the rheology of flocculated suspensions is described and a distinction is made between weakly and strongly flocculated systems. The various semiempirical models that can be applied for the analysis of the flow curves are described.

References

1 Tadros, T. (2005) *Applied Surfactants*, Wiley-VCH Verlag GmbH, Weinheim, Germany.
2 Parfitt, G.D. (ed.) (1977) *Dispersion of Powders in Liquids*, Applied Science Publishers, London.
3 Tadros, Th.F. (ed.) (1987) *Solid/Liquid Dispersions*, Academic Press, London.
4 Gibbs, J.W. (1928) *Collected Works*, vol. 1, Longman, New York, p. 219.
5 Hough, D.B., and Randall, H.M. (1983) *Adsorption from Solution at the Solid/Liquid Interface* (eds G.D. Parfitt and C.H. Rochester), Academic Press, London, p. 247.
6 Fuerstenau, D.W., and Healy, T.W. (1972) *Adsorptive Bubble Separation Techniques* (ed. R. Lemlich), Academic Press, London, p. 91.
7 Somasundaran, P., and Goddard, E.D. (1979) *Mod. Aspects Electrochem.*, **13**, 207.
8 Deryaguin, B.V., and Landau, L. (1941) *Acta Physicochem. USSR*, **14**, 633.
9 Verwey, E.J.W., and Overbeek, J.Th.G. (1948) *Theory of Stability of Lyophobic Colloids*, Elsevier, Amsterdam.
10 Napper, D.H. (1983) *Polymeric Stabilisation of Dispersions*, Academic Press, London.
11 Thompson, W. (Lord Kelvin) (1871). *Phil. Mag.*, **42**, 448.
12 Ottewill, R.H. (1982) *Concentrated Dispersions* (ed. J.W. Goodwin), Royal Society of Chemistry, London. Chapter 9.
13 Ottewill, R.H. (1983) Concentrated dispersions, in *Fundamental Considerations of Science and Technology of Polymer Colloids*, vol. II (eds G.W. Poehlin, R.H. Ottewill, and J.W. Goodwin), Martinus Nijhoff, Boston/The Hague, p. 503.

2
Fundamentals of Wetting and Spreading

2.1
Introduction

Wetting of powders is an important prerequisite for dispersion of powders in liquids, that is, preparation of suspensions. It is essential to wet both the external and internal surfaces of the powder aggregated and agglomerates. As mentioned in Chapter 1, suspensions are applied in many industries such as paints, dyestuffs, printing inks, agrochemicals, pharmaceuticals, paper coatings, detergents, etc.

In all the above processes one has to consider both the equilibrium and dynamic aspects of the wetting process. The equilibrium aspects of wetting can be studied at a fundamental level using interfacial thermodynamics. Under equilibrium, a drop of a liquid on a substrate produces a contact angle θ, which is the angle formed between planes tangent to the surfaces of the solid and liquid at the wetting perimeter. This is illustrated in Figure 2.1, which shows the profile of a liquid drop on a flat solid substrate. An equilibrium between vapor, liquid, and solid is established with a contact angle θ (that is lower than 90°).

The wetting perimeter is frequently referred to as the three-phase line (solid/liquid/vapor); the most common name is the wetting line.

Most equilibrium wetting studies center around measurements of the contact angle. The smaller the angle the better the liquid is said to wet the solid. Typical examples are given in Table 2.1 for water with a surface tension of $72\,mN\,m^{-1}$ on various substrates.

The above values can be roughly used as a measure of wetting of the substrate by water (glass being completely wetted and PTFE very difficult to wet).

The dynamic process of wetting is usually described in terms of a moving wetting line which results in contact angles that change with the wetting velocity. The same name is sometimes given to contact angles that change with time.

Wetting of a porous substrate may also be considered as a dynamic phenomenon. The liquid penetrates through the pores and gives different contact angles depending on the complexity of the porous structure. The study of the wetting of porous substrates is very difficult. The same applies for wetting of agglomerates and aggregates of powders. However, even measurements of apparent contact

Dispersion of Powders in Liquids and Stabilization of Suspensions, First Edition. Tharwat F. Tadros.
© 2012 Wiley-VCH Verlag GmbH & Co. KGaA. Published 2012 by Wiley-VCH Verlag GmbH & Co. KGaA.

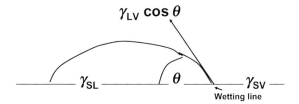

Figure 2.1 Schematic representation of the contact angle and wetting line.

Table 2.1 Typical contact angle values for a water drop on various substrates.

Substrate	Contact angle θ (deg)
PTFE (Teflon)	112
Paraffin wax	110
Polyethylene	103
Human skin	75–90
Glass	0

angles can be very useful for comparing one porous substrate with another and one powder with another.

Despite the increasing attention on the dynamics of wetting, understanding the kinetics of the process at a fundamental level has never been achieved.

The spreading of liquids on substrates is also an important industrial phenomenon. A useful concept introduced by Harkins [1] is the spreading coefficient which is simply the work in destroying a unit area of the solid/liquid and liquid/vapor interface to produce an area of the solid/air interface. The spreading coefficient is simply determined from the contact angle θ and the liquid/vapor surface tension γ_{LV}:

$$S = \gamma_{LV}(\cos\theta - 1) \tag{2.1}$$

For spontaneous spreading S has to be zero or positive. If S is negative only limited spreading is obtained.

2.2
The Concept of the Contact Angle

Wetting is a fundamental interfacial phenomenon in which one fluid phase is displaced completely or partially by another fluid phase from the surface of a solid or a liquid. The most useful parameter that may describe wetting is the contact angle of a liquid on the substrate and this is discussed below.

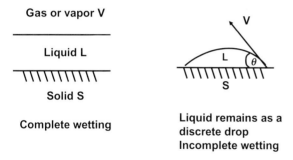

Figure 2.2 Illustration of complete and partial wetting.

2.2.1
The Contact Angle

When a drop of a liquid is placed on a solid, the liquid either spreads to form a thin (uniform) film or remains as a discrete drop. This is schematically illustrated in Figure 2.2.

2.2.2
Wetting Line – Three-Phase Line (Solid/Liquid/Vapor)

The contact angle θ is the angle formed between planes tangent to the surfaces of the solid and liquid at the wetting perimeter. The wetting perimeter is referred to as the three-phase line (solid/liquid/vapor) or simply the wetting line. The utility of contact angle measurements depends on equilibrium thermodynamic arguments (static measurements).

In practical systems, one has to displace fluid (air) with another (liquid) as quickly and as efficiently as possible. Dynamic contact angle measurements (associated with moving wetting line) are more relevant in many practical applications.

Even under static conditions, contact angle measurements are far from being simple since they are mostly accompanied by hysteresis. The value of θ depends on the history of the system and whether the liquid is tending to advance across or recedes from the solid surface. The limiting angles achieved just prior to movement of the wetting line (or just after movement ceases) are known as the advancing and receding contact angles, θ_A and θ_R, respectively. For a given system $\theta_A > \theta_R$ and θ can usually take any value between these two limits without discernible movement of the wetting line.

2.2.3
Thermodynamic Treatment – Young's Equation [2]

The liquid drop takes the shape that minimizes the free energy of the system. Consider a simple system of a liquid drop (L) on a solid surface (S) in equilibrium with the vapor of the liquid (V) as is illustrated in Figure 2.1.

The sum ($\gamma_{SV} A_{SV} + \gamma_{SL} A_{SL} + \gamma_{LV} A_{LV}$) should be a minimum at equilibrium and this leads to Young's equation

$$\gamma_{SV} = \gamma_{SL} + \gamma_{LV} \cos\theta \qquad (2.2)$$

In the above equation θ is the equilibrium contact angle. The angle that a drop assumes on a solid surface is the result of the balance between the cohesion force in the liquid and the adhesion force between the liquid and solid, that is,

$$\gamma_{LV} \cos\theta = \gamma_{SV} - \gamma_{SL} \qquad (2.3)$$

or

$$\cos\theta = \frac{\gamma_{SV} - \gamma_{SL}}{\gamma_{LV}} \qquad (2.4)$$

If there is no interaction between the solid and liquid, then

$$\gamma_{SL} = \gamma_{SV} + \gamma_{LV} \qquad (2.5)$$

that is, $\theta = 180°$ ($\cos\theta = -1$).

If there is strong interaction between solid and liquid (maximum wetting), the latter spreads until Young's equation is satisfied ($\theta = 0$) and

$$\gamma_{LV} = \gamma_{SV} - \gamma_{SL} \qquad (2.6)$$

The liquid spreads spontaneously on the solid surface.

When the surface of the solid is in equilibrium with the liquid vapor, then one must consider the spreading pressure, π_e. As a result of the adsorption of the vapor on the solid surface, its surface tension γ_s is reduced by π_e, that is,

$$\gamma_{SV} = \gamma_s - \pi_e \qquad (2.7)$$

and Young's equation can be written as

$$\gamma_{LV} \cos\theta = \gamma_s - \gamma_{SL} - \pi_e \qquad (2.8)$$

In general, Young's equation provides a precise thermodynamic definition of the contact angle. However, it suffers from the lack of direct experimental verification since both γ_{SV} and γ_{SL} cannot be directly measured. An important criterion for the application of Young's equation is to have a common tangent at the wetting line between the two interfaces.

2.3
Adhesion Tension

There is no direct way by which γ_{SV} or γ_{SL} can be measured. The difference between γ_{SV} and γ_{SL} can be obtained from contact angle measurements. This difference is referred to as the "wetting tension" or "adhesion tension,"

$$\text{Adhesion tension} = \gamma_{SV} - \gamma_{SL} = \gamma_{LV} \cos\theta \qquad (2.9)$$

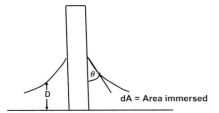

Figure 2.3 Schematic representation of the immersion of a solid plate in a liquid.

Consider the immersion of a solid in a liquid as is illustrated in Figure 2.3. When the plate is immersed in the liquid, an area dA γ_{SV} is lost and an area dA γ_{SL} is formed.

The (Helmholtz) free energy change dF is given by

$$dF = dA(\gamma_{SV} - \gamma_{SL}) \tag{2.10}$$

This is balanced by the force on the plate $W\,dD$:

$$W\,dD = dA(\gamma_{SV} - \gamma_{SL}) \tag{2.11}$$

or

$$W\left(\frac{dD}{dA}\right) = (\gamma_{SV} - \gamma_{SL}) = \gamma_{LV}\cos\theta \tag{2.12}$$

$dA/dD = p$ = plate perimeter

or

$$\frac{W}{p} = \gamma_{LV}\cos\theta \tag{2.13}$$

Equation (2.13) forms the basis of measuring the contact angle θ using an immersed plate (Wilhelmy plate). Equation (2.13) is also the basis of measuring the surface tension of a liquid using the Wilhelmy plate technique. If the plate is made to wet the liquid completely, that is, $\theta = 0$ or $\cos\theta = 1$, then $W/p = \gamma_{LV}$. By measuring the weight of the plate as it touches the liquid one obtains γ_{LV}.

Gibbs [3] defined the adhesion tension τ as the difference between the surface pressure at the solid/liquid interface π_{SL} and that at the solid/vapor interface π_{SV},

$$\tau = \pi_{SL} - \pi_{SV} \tag{2.14}$$

$$\pi_{SL} = \gamma_s - \gamma_{SL} \tag{2.15}$$

$$\pi_{SV} = \gamma_s - \gamma_{SV} \tag{2.16}$$

Combining Eqs. (2.14)–(2.16) with Young's equation, Gibbs arrived at the following equation for the adhesion tension:

$$\tau = \gamma_{SV} - \gamma_{SL} = \gamma_{LV}\cos\theta \tag{2.17}$$

which is identical to Eq. (2.9).

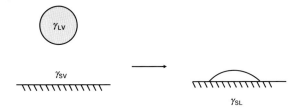

Figure 2.4 Representation of adhesion of a drop on a solid substrate.

Thus, the adhesion tension depends on the measurable quantities γ_{LV} and θ. As long as $\theta < 90°$, the adhesion tension is positive.

2.4
Work of Adhesion W_a

Consider a liquid drop with surface tension γ_{LV} and a solid surface with surface tension γ_{SV}. When the liquid drop adheres to the solid surface it forms a surface tension γ_{SL}. This is schematically illustrated in Figure 2.4.

The work of adhesion [4, 5] is simply the difference between the surface tensions of the liquid/vapor and solid/vapor and that of the solid/liquid:

$$W_a = \gamma_{SV} + \gamma_{LV} - \gamma_{SL} \tag{2.18}$$

Using Young's equation

$$W_a = \gamma_{LV}(\cos\theta + 1) \tag{2.19}$$

2.5
Work of Cohesion

The work of cohesion W_c is the work of adhesion when the two phases are the same. Consider a liquid cylinder with unit cross-sectional area. When this liquid is subdivided into two cylinders, as is illustrated in Figure 2.5, two new surfaces are formed.

The two new areas will have a surface tension of $2\gamma_{LV}$ and the work of cohesion is simply

$$W_c = 2\gamma_{LV} \tag{2.20}$$

Thus, the work of cohesion is simply equal to twice the liquid surface tension. An important conclusion may be drawn if one considers the work of adhesion given by Eq. (2.19) and the work of cohesion given by Eq. (2.20): when $W_c = W_a$, $\theta = 0°$. This is the condition for complete wetting. When $W_c = 2W_a$, $\theta = 90°$ and the liquid forms a discrete drop on the substrate surface.

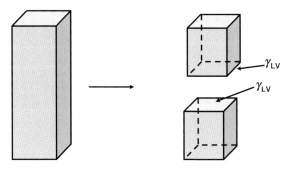

Figure 2.5 Schematic representation of subdivision of a liquid cylinder.

Thus, the competition between the cohesion of the liquid to itself and its adhesion to a solid gives an angle of contact that is constant and specific to a given system at equilibrium. This shows the importance of Young's equation in defining wetting.

2.6
Calculation of Surface Tension and Contact Angle

Fowler [6] was the first to calculate the surface tension of a simple liquid. The basic idea is to use the intermolecular forces that operate between atoms and molecules. For this purpose it is sufficient to consider the various van der Waals forces: dipole–dipole (Keesom)–dipole–induced dipole (Debye)–Dispersion (London).

Dispersion forces are the most important since they occur between all atoms and molecules and they are additive. The London expression for the dispersion interaction u between two molecules separated by a distance r is

$$u = -\frac{\beta}{r^6} \quad (2.21)$$

where β is the London dispersion constant (that depends on the electric polarizability of the molecules).

Hamaker [7] calculated the attractive forces between macroscopic bodies using a simple additivity principle. For two semi-infinite flat plates separated by a distance d, the attractive force F is given by

$$F = \frac{A}{6\pi d^3} \quad (2.22)$$

where A is the Hamaker constant that is given by

$$A = \pi^2 n^2 \beta \quad (2.23)$$

where n is the number of interacting dispersion centers per unit volume.

Fowler [6] used the above intermolecular theory to calculate the energy required to break a column of liquid of unit cross section and remove the two halves to infinite separation. This gave the work of cohesion and to be equal to twice the surface tension. He used statistical thermodynamic methods to integrate the intermolecular forces. These were calculated from the Lennard–Jones equation, which is like the London equation (2.21), but contains an additional term which varies as r^{-12} to account for very short range intermolecular repulsion.

2.6.1
Good and Girifalco Approach [8, 9]

Good and Girifalco [8, 9] proposed a more empirical approach to the problem of calculating the surface and interfacial tension. The interaction constant for two different particles was assumed to be equal to the geometric mean of the interaction constants for the individual particles. This is referred to as the Berthelot principle.

For two atoms i and j with London constants β_i and β_j, the interaction constant β_{ij} is given by the expression

$$\beta_{ij} = (\beta_i \beta_j)^{1/2} \qquad (2.24)$$

Similarly the Hamaker constant A_{ij} is given by the geometric mean of the individual Hamaker constants:

$$A_{ij} = (A_i A_j)^{1/2} \qquad (2.25)$$

By analogy Good and Girifalco [8] represented the work of adhesion between two different liquids W_{a12} as the geometric mean of their respective works of cohesion:

$$W_{a12} = \varphi(W_{c1} W_{c2})^{1/2} \qquad (2.26)$$

where ϕ is a constant that depends on the relative molecular size and polar content of the interacting media.

The interfacial tension γ_{12} is then related to the surface tension of the individual liquids γ_1 and γ_2 by the following expression:

$$\gamma_{12} = \gamma_1 + \gamma_2 - 2\varphi(\gamma_1 \gamma_2)^{1/2} \qquad (2.27)$$

For nonpolar media, Eq. (2.27) was found to work well with $\varphi \sim 1$. For dissimilar substances such as water and alkanes, the value of ϕ ranges from 0.35 to 1.15.

Good and Grifalco [8] and Good [9] extended the above treatment to the solid/liquid interface and they obtained the following expression for the contact angle θ:

$$\cos\theta = -1 + 2\varphi \left(\frac{\gamma_s}{\gamma_1}\right)^{1/2} - \frac{\pi_s}{\gamma_1} \qquad (2.28)$$

where π_s is the surface pressure of fluid 1 adsorbed at the solid/gas interface. Equation (2.28) gave reasonable values for γ_s for nonpolar substrates.

Although the above analysis is semiempirical, it can be usefully applied to predict the interfacial tension between two immiscible liquids. The analysis is also useful for predicting the surface tension of a solid substrate from measurement of the contact angle of the liquid.

2.6.2
Fowkes Treatment [10]

Fowkes [10] proposed that the surface and interfacial tensions can be subdivided into independent additive terms arising from different types of intermolecular interactions. For water in which both hydrogen bonding and dispersion forces operate, the surface tension can be assumed to be the sum of two contributions

$$\gamma = \gamma^h + \gamma^d \tag{2.29}$$

For nonpolar liquids such as alkanes, γ is simply equal to γ^d. By applying the geometric mean relationship to γ^d, Fowkes [10] obtained the following expression for the work of adhesion W_{a12}:

$$W_{a12} = 2(\gamma_1^d \gamma_2^d)^{1/2} \tag{2.30}$$

Thus, the interfacial tension γ_{12} is given by the following expression:

$$\gamma_{12} = \gamma_1 + \gamma_2 - 2(\gamma_1^d \gamma_2^d)^{1/2} \tag{2.31}$$

Fowkes [10] assumed that the nondispersive contributions to γ_1 and γ_2 are unaltered at the 12 interface.

Similar equations can be written for the solid/liquid interface. An expression for the contact angle has been derived by Fowkes [10],

$$\cos\theta = -1 + \frac{2(\gamma_s^d \gamma_1^d)^{1/2}}{\gamma_1} - \frac{\pi_s}{\gamma_1} \tag{2.32}$$

Various studies showed that Eqs (2.31) and (2.32) are quite effective for materials that interact only through dispersion forces and gave reasonable predictions for γ^d for liquids and solids in which other forces are active. The value for γ^d for water is ~21.8 mN m^{-1} leaving γ^h ~ 51 mN m^{-1}. Fowkes [10] showed that for nonpolar liquids (Eq. (2.31)) can be obtained by summation of pairwise interactions.

2.7
The Spreading of Liquids on Surfaces

2.7.1
The Spreading Coefficient S

Harkins [11, 12] defined the initial spreading coefficient as the work required to destroy the unit area of solid/liquid (SL) and liquid/vapor (LV) and leaves the unit area of bare solid (SV). This is schematically represented in Figure 2.6.

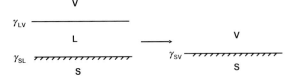

Figure 2.6 Schematic representation of the spreading coefficient S.

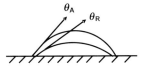

Figure 2.7 Schematic representation of advancing and receding angles.

$$S = \gamma_{SV} - (\gamma_{SL} + \gamma_{LV}) \tag{2.33a}$$

Using Young's equation

$$S = \gamma_{LV}(\cos\theta + 1) \tag{2.33b}$$

If S is positive, the liquid will spread until it completely wets the solid so that $\theta = 0°$. If S is negative ($\theta > 0°$) only partial wetting occurs. Alternatively one can use the equilibrium or final spreading coefficient.

2.8
Contact Angle Hysteresis

For a liquid spreading on a uniform nondeformable solid (idealized case), there is only one contact angle (the equilibrium value). With real surfaces (practical systems) a number of stable angles can be measured. Two relatively reproducible angles can be measured: largest, advancing angle θ_A and smallest, receding angle θ_R. This is illustrated in Figure 2.7.

θ_A is measured by advancing the periphery of the drop over the surface (e.g., by adding more liquid to the drop). θ_R is measured by pulling the liquid back (e.g., by removing some liquid from the drop). The difference between θ_A and θ_{3R} is termed "contact angle hysteresis." The contact angle hysteresis can be illustrated by placing a drop on a tilted surface with an angle α from the horizontal as is illustrated in Figure 2.8.

The advancing and receding angle are clearly shown at the front and the back of the drop on the tilted surface. Due to the gravity field ($mg \sin \alpha \, dl$, m = mass of the drop, g = acceleration due to gravity), the drop will slide until the difference between the work of dewetting and wetting balances the gravity force:

2.8 Contact Angle Hysteresis

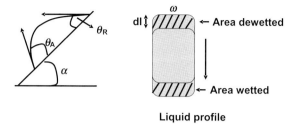

Figure 2.8 Representation of a drop profile on a tilted surface.

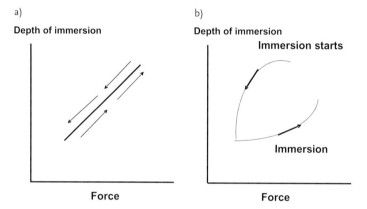

Figure 2.9 Relationship between depth of immersion and force: (a) no hysteresis; (b) hysteresis present.

$$\text{Work of dewetting} = \gamma_{LV}(\cos\theta_R + 1)\omega\, dl \qquad (2.34)$$

$$\text{Work of wetting} = \gamma_{LV}(\cos\theta_A + 1)\omega\, dl \qquad (2.35)$$

$$mg\sin\alpha\, dl = \gamma_{LV}(\cos\theta_R - \cos\theta_A)\omega\, dl \qquad (2.36)$$

$$\frac{mg\sin\alpha}{\omega} = \gamma_{LV}(\cos\theta_R - \cos\theta_A) \qquad (2.37)$$

Hysteresis can be demonstrated by measuring the force on a plate that is continuously immersed in the liquid. When the plate is immersed, the force will decrease due to buoyancy. When there is no contact angle hysteresis, the relationship between the depth of immersion and force will be as shown in Figure 2.9a. With hysteresis, the relationship between the depth of immersion and force will be as shown in Figure 2.9b.

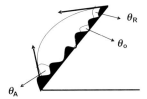

Figure 2.10 Representation of a drop profile on a rough surface.

2.8.1
Reasons for Hysteresis

i) Penetration of wetting liquid into pores during advancing contact angle measurements.

ii) **Surface roughness**: the first and rear edges meet the solid with the same intrinsic angle θ_o. The macroscopic angles θ_A and θ_R vary significantly at the front and the rear of the drop. This is illustrated in Figure 2.10.

2.8.2
Wenzel's Equation [13]

Wenzel [13] considered the true area of a rough surface A (which takes into account all the surface topography, peaks, and valleys) and the projected area A' (the macroscopic or apparent area).

A roughness factor r can be defined as

$$r = \frac{A}{A'} \tag{2.38}$$

$r > 1$; the higher the value of r the higher the roughness of the surface.

The measured contact angle θ (the macroscopic angle) can be related to the intrinsic contact angle θ_o through r

$$\cos\theta = r\cos\theta_o \tag{2.39}$$

Using Young's equation

$$\cos\theta = r\left(\frac{\gamma_{SV} - \gamma_{SL}}{\gamma_{LV}}\right) \tag{2.40}$$

If $\cos\theta$ is negative on a smooth surface ($\theta > 90°$), it becomes more negative on a rough surface; θ becomes larger and surface roughness reduces wetting. If $\cos\theta$ is positive on a smooth surface ($\theta < 90°$), it becomes more positive on a rough surface; θ is smaller and surface roughness enhances wetting.

Surface heterogeneity: most real surfaces are heterogeneous consisting of patches (islands) that vary in their degrees of hydrophilicity/hydrophobicity. As the drop advances on such a heterogeneous surface, the edge of the drop tends to stop at the boundary of the island. The advancing angle will be associated with the intrinsic angle of the high contact angle region (the more hydrophobic patches or islands). The receding angle will be associated with the low contact angle region, that is, the more hydrophilic patches or islands.

If the heterogeneities are small compared with the dimensions of the liquid drop, one can define a composite contact angle. Cassie [14, 15] considered the maximum and minimum values of the contact angles and used the following simple expression:

$$\cos\theta = Q_1 \cos\theta_1 + Q_2 \cos\theta_2 \tag{2.41}$$

Q_1 is the fraction of surface having contact angle θ_1 and Q_2 is the fraction of surface having contact angle θ_2. θ_1 and θ_2 are the maximum and minimum contact angles, respectively.

References

1 Blake, T.B. (1984) Wetting, in *Surfactants* (ed. Th.F. Tadros), Academic Press, London, pp. 221–275.
2 Young, T. (1805) *Phil. Trans. R. Soc. Lond.*, **95**, 65.
3 Gibbs, J.W. (1928) *The Collected Work of J. Willard Gibbs*, vol. 1, Longman-Green, New York.
4 Everett, D.H. (1980) *Pure Appl. Chem.*, **52**, 1279.
5 Johnson, R.E. (1959) *J. Phys. Chem.*, **63**, 1655.
6 Fowler, R.H. (1937) *Proc. R. Soc. A*, **159**, 229.
7 Hamaker, H.C. (1937) *Physica*, **4**, 1058.
8 Good, R.J., and Girifalco, L.A. (1960) *J. Phys. Chem.*, **64**, 561.
9 Good, R.J. (1964) *Adv. Chem. Ser.*, **43**, 74.
10 Fowkes, F.M. (1964) *Adv. Chem. Ser.*, **43**, 99.
11 Harkins, W.D. (1952) *The Physical Chemistry of Surface Films*, Reinhold, New York.
12 Harkins, W.D. (1937) *J. Phys. Chem.*, **5**, 135.
13 Wenzel, R.N. (1936) *Ind. Eng. Chem.*, **28**, 988.
14 Cassie, A.B.D., and Dexter, S. (1944) *Trans. Faraday Soc.*, **40**, 546.
15 Cassie, A.B.D. (1948) *Discuss. Faraday Soc.*, **3**, 361.

3
The Critical Surface Tension of Wetting and the Role of Surfactants in Powder Wetting

3.1
The Critical Surface Tension of Wetting

A systematic way of characterizing "wettability" of a surface was introduced by Zisman [1]. The contact angle exhibited by a liquid on a low-energy surface is largely dependent on the surface tension of the liquid γ_{LV}. For a given substrate and a series of related liquids (such as n-alkanes, siloxanes, or dialkyl ethers) $\cos\theta$ is a linear function of the liquid surface tension γ_{LV}. This is illustrated in Figure 3.1 for a number of related liquids on polytetrafluoroethylene (PTFE). The figure also shows the results for unrelated liquids with widely ranging surface tensions; the line broadens into a band which tends to be curved for high surface tension polar liquids.

The surface tension at the point where the line cuts the $\cos\theta = 1$ axis is known as the critical surface tension of wetting. γ_c is the surface tension of a liquid that would just spread on the substrate to give complete wetting. If $\gamma_{LV} \leq \gamma_c$ the liquid will spread, whereas for $\gamma_{LV} > \gamma_c$ the liquid will form a nonzero contact angle.

The above linear relationship can be represented by the following empirical equation:

$$\cos\theta = 1 - b(\gamma_{LV} - \gamma_c) \tag{3.1}$$

where the constant b usually has a value between 0.03 and 0.04.

High-energy solids such as glass and polyethylene terphthalate have high critical surface tension ($\gamma_c > 40\,\text{mN m}^{-1}$). Lower energy solids such as polyethylene have lower values of γ_c ($\sim 31\,\text{mN m}^{-1}$). The same applies to hydrocarbon surfaces such as paraffin wax. Very low energy solids such as PTFE have lower γ_c of the order of $18\,\text{mN m}^{-1}$. The lowest known value is $\sim 6\,\text{mN m}^{-1}$ which is obtained using condensed monolayers of perfluorolauric acid.

The measurement of γ_c requires a range of liquids with widely varying surface tension which is difficult to obtain. Alternatively, one can use a series of aqueous surfactant solutions with varying concentrations or binary mixtures of liquids such as alcohols, glycols, and water. Unfortunately, using surfactant solutions with different concentrations can give wrong values of γ_c due to surfactant adsorption on the solid surface thus altering the surface characteristics.

Dispersion of Powders in Liquids and Stabilization of Suspensions, First Edition. Tharwat F. Tadros.
© 2012 Wiley-VCH Verlag GmbH & Co. KGaA. Published 2012 by Wiley-VCH Verlag GmbH & Co. KGaA.

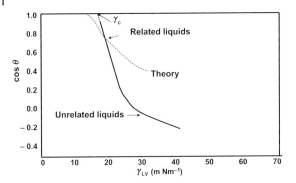

Figure 3.1 Variation of $\cos\theta$ with γ_{LV} for related and unrelated liquids on PTFE.

3.2
Theoretical Basis of the Critical Surface Tension

The value of γ_c depends to some extent on the set of liquids used to measure it. Zisman [1] described γ_c as "a useful empirical parameter" whose relative values act as one would expect of the specific surface free energy of the solid, γ_s^0.

Several authors were tempted to identify γ_c with γ_s (the surface tension of the solid substrate) or γ_1^d (the dispersion component of the surface tension). Good and Girifalco [2, 3] suggested the following expression for the contact angle:

$$\cos\theta = -1 + 2\varphi\left(\frac{\gamma_s}{\gamma_{LV}}\right)^{1/2} - \frac{\pi_{SV}}{\gamma_{LV}} \tag{3.2}$$

where ϕ is an empirical constant and π_{SV} is the surface pressure of the liquid vapor adsorbed at the solid/liquid interface.

With $\pi_{SV} = 0$ and $\cos\theta = 1$,

$$\gamma_{SL} = \gamma_{LV} = \varphi^2 \gamma_s = \gamma_c \tag{3.3}$$

For nonpolar liquids and solids $\phi \sim 1$ and $\gamma_s \sim \gamma_c$.

Fowkes [4] obtained the following equation for the contact angle of a liquid on a solid substrate:

$$\cos\theta = -1 + \frac{2(\gamma_s^d \gamma_{LV}^d)^{1/2}}{\gamma_{LV}} - \frac{\pi_{SV}}{\gamma_{LV}} \tag{3.4}$$

Again putting $\cos\theta = 1$ and $\pi_{SV} = 0$,

$$\gamma_{SL} = \gamma_{LV} = (\gamma_{LV}^d \gamma_s^d)^{1/2} = \gamma_c \tag{3.5}$$

Equations (3.2) and (3.4) predict that if $\pi_{SV} = 0$, a plot of $\cos\theta$ versus γ_{LV} should give a straight line with intercept $(\gamma_c)^{-1/2}$ on the $\cos\theta = 1$ axis. The experimental results seem to support this prediction. Thus, for nonpolar solids, $\gamma_c = \gamma_s$, provided $\pi_{SV} = 0$, that is, there is no adsorption of liquid vapor on the substrate. The above condition is unlikely to be satisfied when $\theta = 0$.

3.3 Effect of Surfactant Adsorption

Surfactants lower the surface tension of the liquid, γ_{LV}, and they also adsorb at the solid/liquid interface lowering γ_{SL}. The adsorption of surfactants at the liquid/air interface can be easily described by the Gibbs adsorption equation [5],

$$\frac{d\gamma_{LV}}{dC} = -2.303 \Gamma RT \qquad (3.6)$$

where C is the surfactant concentration (mol dm^{-3}) and Γ is the surface excess (amount of adsorption in mol m^{-2}).

Γ can be obtained from surface tension measurements using solutions with various molar concentrations (C). From a plot of γ_{LV} versus log C one can obtain Γ from the slope of the linear portion of the curve just below the critical micelle concentration (cmc).

The adsorption of surfactant at the solid/liquid interface also lowers γ_{SL}. From Young's equation,

$$\cos\theta = \frac{\gamma_{SV} - \gamma_{SL}}{\gamma_{LV}} \qquad (3.7)$$

Surfactants reduce θ if either γ_{SL} or γ_{LV} or both are reduced (when γ_{SV} remains constant). Smolders [6] obtained an equation for the change of contact angle with surfactant concentration by differentiating Young's equation with respect to ln C at constant temperature,

$$\frac{d(\gamma_{LV}\cos\theta)}{d\ln C} = \frac{d\gamma_{SV}}{d\ln C} - \frac{d\gamma_{SL}}{d\ln C} \qquad (3.8)$$

Using the Gibbs equation,

$$\sin\theta\left(\frac{d\theta}{dlC}\right) = RT(\Gamma_{SV} - \Gamma_{SL} - \Gamma_{LV}\cos\theta) \qquad (3.9)$$

Since $\gamma_{LV} \sin\theta$ is always positive, $(d\theta/d\ln C)$ will always have the same sign as the right-hand side of Eq. (3.9) and three cases may be distinguished:

i) $(d\theta/d\ln C) < 0 - \Gamma_{SV} < \Gamma_{SL} + \Gamma_{LV}\cos\theta$. Addition of surfactant improves wetting.
ii) $(d\theta/d\ln C) = 0 - \Gamma_{SV} = \Gamma_{SL} + \Gamma_{LV}\cos\theta$ (no effect).
iii) $(d\theta/d\ln C) > 0 - \Gamma_{SV} > \Gamma_{SL} + \Gamma_{LV}\cos\theta$. Addition of surfactant causes dewetting.

In many practical situations, the mode of behavior may vary with surfactant concentration, for example, from dewetting to wetting. This is particularly the case with polar surfaces such as silica. On addition of a cationic surfactant to a negatively charged silica surface (at pH > 4, that is, above its isoelectric point of ~2–3) dewetting occurs at low surfactant concentration due to the electrostatic attraction between the cationic surfactant head group with the negative charges on the surface with the alky chains pointing toward the solution and the surface becomes more hydrophobic resulting in dewetting. With the increase in cationic surfactant concentration a bilayer of surfactant molecules is formed by hydrophobic interaction between the alky chains leaving the polar positively charged head group pointing toward the solution and wetting occurs.

3.4
Dynamic Processes of Adsorption and Wetting

In the process of wetting of powders one should consider the dynamic process of surfactant adsorption both at the liquid/vapor (LV) and solid/liquid (SL) interfaces. The most frequently used parameter to characterize the dynamic process of adsorption at the LV interface is the dynamic surface tension (that is time-dependent quantity). Techniques should be available to measure γ_{LV} as a function of time (ranging from a fraction of a millisecond to minutes and hours or days). At the SL interface one should also consider the rate of adsorption. To optimize the use of surfactants, polymers, and mixtures of them specific knowledge of their dynamic adsorption behavior rather than equilibrium properties is of great interest [7]. It is, therefore, necessary to describe the dynamics of surfactant adsorption at a fundamental level.

3.4.1
General Theory of Adsorption Kinetics

The first physically sound model for adsorption kinetics was derived by Ward and Tordai [8]. It is based on the assumption that the time dependence of surface or interfacial tension, which is directly proportional to the surface excess Γ (mol m^{-2}), is caused by diffusion and transport of surfactant molecules to the interface. This is referred to as "the diffusion controlled adsorption kinetics model." The interfacial surfactant concentration at any time t, $\Gamma(t)$, is given by the following expression:

$$\Gamma(t) = 2\left(\frac{D}{\pi}\right)^{1/2} \left(c_0 t^{1/2} - \int_0^{t^{1/2}} c(0, t-\tau) \, d(\tau)^{1/2}\right) \tag{3.10}$$

where D is the diffusion coefficient, c_0 is the bulk concentration, and τ is the thickness of the diffusion layer.

The above diffusion-controlled model assumes transport by diffusion of the surface active molecules to be the rate-controlled step. The so-called kinetic controlled model is based on the transfer mechanism of molecules from solution to the adsorbed state and vice versa [7].

A schematic picture of the interfacial region is given in Figure 3.2 which shows three main states: (i) adsorption when the surface concentration Γ is lower than the equilibrium value Γ_0; (ii) equilibrium state when $\Gamma = \Gamma_0$, and (iii) desorption when $\Gamma > \Gamma_0$.

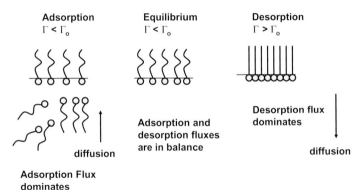

Figure 3.2 Representation of the fluxes of adsorbed surfactant molecules in the absence of liquid flow.

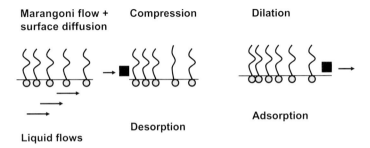

Figure 3.3 Representation of surfactant transport at the surface and in the bulk of a liquid.

The transport of surfactant molecules from the liquid layer adjacent to the interface (subsurface) is simply determined by molecular movements (in the absence of forced liquid flow). At equilibrium, that is, when $\Gamma = \Gamma_0$, the flux of adsorption is equal to the flux of desorption. Clearly when $\Gamma < \Gamma_0$, the flux of adsorption predominates, whereas when $\Gamma > \Gamma_0$, the flux of desorption predominates [7].

In the presence of liquid flow, the situation becomes more complicated due to the creation of surface concentration gradients [7]. These gradients, described by the Gibbs dilational elasticity [9], initiate a flow of mass along the interface in direction of to the higher surface or interfacial tension (Marangoni effect). This situation can happen, for example, if an adsorption layer is compressed or stretched, as is illustrated in Figure 3.3.

A qualitative model that can describe adsorption kinetics is described by the following equation:

$$\Gamma(t) = c_0 \left(\frac{Dt}{\pi} \right)^{1/2} \tag{3.11a}$$

Equation (3.11) gives a rough estimate and results from Eq. (3.10) when the second term on the right-hand side is neglected.

An equivalent equation to (3.11) has been derived by Panaitov and Petrov [9],

$$c(0,t) = c_0 - \frac{2}{(D\pi)^{1/2}} \int_0^{t^{1/2}} \frac{d\Gamma(t-\tau)}{dt} d\tau^{1/2} \tag{3.11b}$$

Hansen [10], and Miller and Lunkenheimer [11] gave numerical solutions to the integrals of Eqs. (3.10) and (3.12) and obtained a simple expression using a Langmuir isotherm,

$$\Gamma(t) = \Gamma_\infty \frac{c(0,t)}{a_L + c(0,t)} \tag{3.12}$$

where a_L is the constant in the Langmuir isotherm (mol m^{-3})

The corresponding equation for the variation of surface tension γ with time is as follows (Langmuir–Szyszowski equation),

$$\gamma = \gamma_0 + RT\Gamma_\infty \ln\left(1 - \frac{\Gamma(t)}{\Gamma_\infty}\right) \tag{3.13}$$

Calculation based on Eqs. (3.12)–(3.14) are given in Figure 3.4, with different values of c_0/a_L [7].

3.4.2
Adsorption Kinetics from Micellar Solutions

Surfactants form micelles above the cmc of different sizes and shapes, depending on the nature of the molecule, temperature, electrolyte concentration, etc. The dynamic nature of micellization can be described by two main relaxation processes, τ_1 (the lifetime of a monomer in a micelle) and τ_2 (the lifetime of the micelle, that is, complete dissolution into monomers).

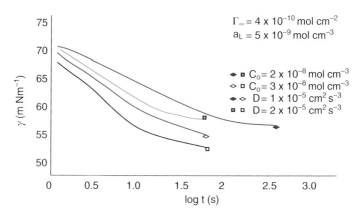

Figure 3.4 Surface tension γ–log t curves calculated on the basis of Eqs. (3.11)–(3.13).

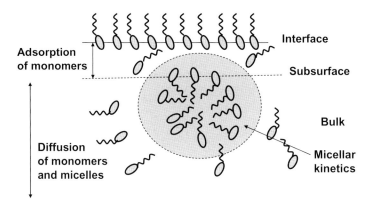

Figure 3.5 Representation of the adsorption process from a micellar solution.

The presence of micelles in equilibrium with monomers influences the adsorption kinetics remarkably. After a fresh surface has been formed surfactant monomers are adsorbed resulting in a concentration gradient of these monomers. This gradient will be equalized by diffusion to re-establish a homogeneous distribution. Simultaneously, the micelles are no longer in equilibrium with monomers within the range of concentration gradient. This leads to a net process of micelle dissolution or rearrangement to re-establish the local equilibrium. As a consequence, a concentration gradient of micelles results, which is equalized by diffusion of micelles [7].

Based on the above concepts, one would expect that the ratio of monomers c_1 to micelles c_m, the aggregation number n, rate of micelle formation k_f, and micelle dissolution k_d will influence the rate of the adsorption process. A schematic picture of the kinetic process in the presence of micelles is given in Figure 3.5.

The above picture shows that to describe the kinetics of adsorption, one must take into account the diffusion of monomers and micelles as well as the kinetics of micelle formation and dissolution. Several processes may take place and these are represented schematically in Figure 3.6. Three main mechanisms may be considered, namely formation dissolution (Figure 3.6a), rearrangement (Figure 3.6b), and stepwise aggregation dissolution (Figure 3.6c). To describe the effect of micelles on adsorption kinetics, one should know several parameters such as micelle aggregation number and rate constants of micelle kinetics [12].

3.4.3
Experimental Techniques for Studying Adsorption Kinetics

The two most suitable techniques for studying adsorption kinetics are the drop volume method and the maximum bubble pressure method. The first method can obtain information on adsorption kinetics in the range of seconds to some minutes. The maximum bubble pressure method allows one to obtain measurements in the millisecond range. Below a description of both techniques is given.

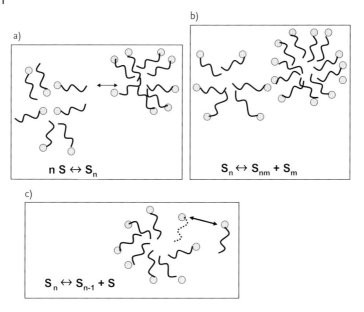

Figure 3.6 Scheme of micelle kinetics.

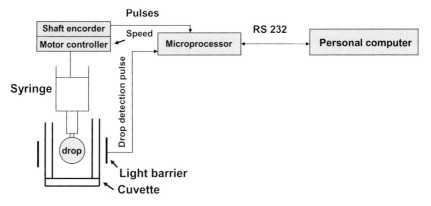

Figure 3.7 Representation of the drop-volume apparatus.

3.4.3.1 The Drop Volume Technique

A schematic representation of the drop volume apparatus [13] is given in Figure 3.7. A metering system in the form of a motor-driven syringe allows the formation of the liquid drop at the tip of a capillary, which is positioned in a sealed cuvette. The cuvette is filled either with a small amount of the measuring liquid, to saturate the atmosphere, or with a second liquid in the case of interfacial studies. A light barrier arranged below the forming drop enables the detection of drop detachment from the capillary. Both the syringe and the light barriers are computer controlled and allow a fully automatic operation of the setup. The syringe and the cuvette are temperature controlled by a water jacket which makes interfacial tension measurements possible in the temperature range 10–90 °C.

As mentioned above, the drop volume method is of dynamic character and it can be used for adsorption processes in the time interval of seconds up to some minutes. At small drop time, the so-called hydrodynamic effect has to be considered [14]. This gives rise to apparently higher surface tension. Kloubek et al. [15] used an empirical equation to account for this effect,

$$V_e = V(t) - \frac{K_v}{t} \tag{3.14}$$

V_e is the unaffected drop volume and $V(t)$ is the measured drop volume. K_v is a proportionality factor that depends on surface tension γ, density difference $\Delta\rho$, and tip radius r_{cap}.

Miller [7] obtained the following equation for the variation of drop volume $V(t)$ with time,

$$V(t) = V_e + t_0 F = V_e \left(1 + \frac{t_0}{t - t_0}\right) \tag{3.15}$$

where F is the liquid flow per unit time that is given by

$$F = \frac{V(t)}{t} = \frac{V_e}{t - t_0} \tag{3.16}$$

The drop volume technique is limited in its application. Under conditions of fast drop formation and larger tip radii, the drop formation shows irregular behavior.

3.4.3.2 Maximum Bubble Pressure Technique

This is the most useful technique for measuring adsorption kinetics at short times, particularly if correction for the so-called dead time, τ_d, is made. The dead time is simply the time required to detach the bubble after it has reached its hemispherical shape. A schematic representation of the principle of maximum bubble pressure is shown in Figure 3.8, which describes the evolution of a bubble at the tip of a capillary. The figure also shows the variation of pressure p in the bubble with time.

At $t = 0$ (initial state), the pressure is low (note that the pressure is equal to $2\gamma/r$; since r of the bubble is large p is small). At $t = \tau$ (smallest bubble radius that is equal to the tube radius) p reaches a maximum. At $t = \tau_b$ (detachment time) p decreases since the bubble radius increases. The design of a maximum bubble pressure method for high bubble formation frequencies (short surface age) requires the following: (i) measurement of bubble pressure; (ii) measurement of bubble formation frequency; and (iii) estimation of surface lifetime and effective surface age. The first problem can be easily solved if the system volume (which is connected to the bubble) is large enough in comparison with the bubble separating from the capillary. In this case, the system pressure is equal to the maximum bubble pressure. The use of an electric pressure transducer for measuring bubble

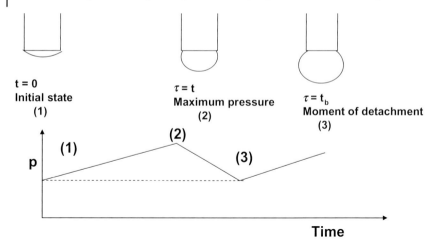

Figure 3.8 Scheme of bubble evolution and pressure change with time.

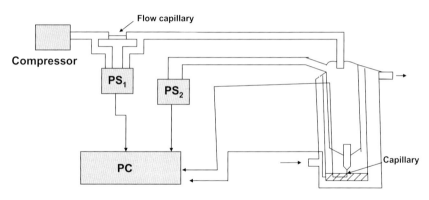

Figure 3.9 Maximum bubble pressure apparatus.

formation frequency presumes that pressure oscillations in the measuring system are distinct enough and this satisfies (ii). Estimation of the surface lifetime and effective surface age, that is, (iii), requires estimation of the dead time τ_d. A schematic representation of the setup for measuring the maximum bubble pressure and surface age is shown in Figure 3.9. The air coming from a microcompressor flows first through the flow capillary. The air flow rate is determined by measuring the pressure difference at both ends of the flow capillary with the electric transducer PS1. Thereafter, the air enters the measuring cell and the excess air pressure in the system is measured by a second electric sensor PS2. In the tube which leads the air to the measuring cell, a sensitive microphone is placed.

The measuring cell is equipped with a water jacket for temperature control, which simultaneously holds the measuring capillary and two platinum electrodes, one of which is immersed in the liquid under study and the second is situated exactly opposite to the capillary and controls the size of the bubble. The electric signals from the gas flow sensor PS1 and pressure transducer PS2, the microphone and the electrodes, as well as the compressor are connected to a personal computer which operates the apparatus and acquires the data.

The value of τ_d, equivalent to the time interval necessary to form a bubble of radius R, can be calculated using Poiseuille's law,

$$\tau_d = \frac{\tau_b L}{Kp}\left(1 + \frac{3r_{ca}}{2R}\right) \qquad (3.17)$$

K is given by Poiseuille's law,

$$K = \frac{\pi r^4}{8\eta l} \qquad (3.18)$$

η is the gas viscosity, l is the length, L is the gas flow rate, and r_{ca} is the radius of the capillary.

The calculation of dead time τ_d can be simplified when taking into account the existence of two gas flow regimes for the gas flow leaving the capillary: bubble flow regime when $\tau > 0$ and jet regime when $\tau = 0$ and hence $\tau_b = \tau_d$. A typical dependence of p on L is shown in Figure 3.10.

On the right-hand side of the critical point the dependence of p on L is linear in accordance with the Poiseuille law. Under these conditions,

$$\tau_d = \tau_b \frac{L p_c}{L_c p} \qquad (3.19)$$

where L_c and p_c are related to the critical point, and L and p are the actual values of the dependence left from the critical point.

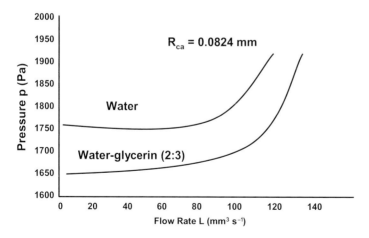

Figure 3.10 Dependence of p on the gas flow rate L at 30 °C.

The surface lifetime can be calculated from

$$\tau = \tau_b - \tau_d = \tau_b\left(1 - \frac{Lp_c}{L_c p}\right) \quad (3.20)$$

The critical point in the dependence of p and L can be easily located and is included in the software of the computer program.

The surface tension value in the maximum bubble pressure method is calculated using the Laplace equation,

$$p = \frac{2\gamma}{r} + \rho h g + \Delta p \quad (3.21)$$

where ρ is the density of the liquid, g is the acceleration due to gravity, h is the depth the capillary is immersed in the liquid, and Δp is a correction factor to allow for hydrodynamic effects.

3.5
Wetting of Powders by Liquids

Wetting of powders by liquids is very important in their dispersion, for example, in the preparation of concentrated suspensions. The particles in a dry powder form either aggregates or agglomerates. In the case of aggregates the particles are joined by their crystal faces. They form compact structures with relatively high bulk density. With agglomerates the particles are joined by their edges or corners and they form loose structures with lower bulk density than those of the aggregates.

It is essential in the dispersion process to wet both external and internal surfaces and displace the air entrapped between the particles. Wetting is achieved by the use of surface active agents (wetting agents) of the ionic or nonionic type which are capable of diffusing quickly (i.e., lower the dynamic surface tension) to the solid/liquid interface and displace the air entrapped by rapid penetration through the channels between the particles and inside any "capillaries." For wetting of hydrophobic powders into water, anionic surfactants, for example, alkyl sulfates or sulfonates or nonionic surfactants of the alcohol or alkyl phenol ethoxylates are usually used (see below).

The process of wetting of a solid by a liquid involves three types of wetting: adhesion wetting, W_a; immersion wetting W_i; and spreading wetting W_s. This can be illustrated by considering a cube of solid with unit area of each side (Figure 3.11).

In every step one can apply Young's equation,

$$\gamma_{SV} = \gamma_{SL} + \gamma_{LV}\cos\theta \quad (3.22)$$

$$W_a = \gamma_{SL} - (\gamma_{SV} + \gamma_{LV}) = -\gamma_{LV}(\cos\theta + 1) \quad (3.23)$$

$$W_i = 4\gamma_{SL} - 4\gamma_{SV} = -4\gamma_{LV}\cos\theta \quad (3.24)$$

$$W_s = (\gamma_{SL} + \gamma_{LV}) - \gamma_{SV} = -\gamma_{LV}(\cos\theta - 1) \quad (3.25)$$

3.5 Wetting of Powders by Liquids

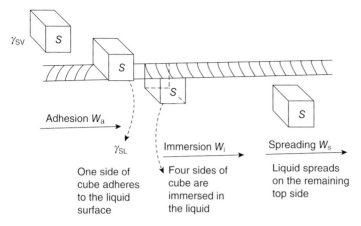

Figure 3.11 Schematic representation of wetting of a cube of solid.

The work of dispersion W_d is the sum of W_a, W_i, and W_s,

$$W_d = W_a + W_i + W_s = 6\gamma_{SV} - \gamma_{SL} = -6\gamma_{LV}\cos\theta \qquad (3.26)$$

Wetting and dispersion depends on the liquid surface tension γ_{LV} and contact angle between liquid and solid, θ. W_a, W_i, and W_s are spontaneous when $\theta < 90°$. W_d is spontaneous when $\theta = 0$. Since surfactants are added in sufficient amounts ($\gamma_{dynamic}$ is lowered sufficiently) spontaneous dispersion is the rule rather than the exception.

Wetting of the internal surface requires penetration of the liquid into channels between and inside the agglomerates. The process is similar to forcing a liquid through fine capillaries. To force a liquid through a capillary with radius r, a pressure p is required that is given by

$$W_d = W_a + W_i + W_s = 6\gamma_{SV} - \gamma_{SL} = -6\gamma_{LV}\cos\theta \qquad (3.27)$$

γ_{SL} has to be made as small as possible; rapid surfactant adsorption to the solid surface, low θ. When $\theta = 0$, $p \propto \gamma_{LV}$. Thus for penetration into pores one requires a high γ_{LV}. Thus, wetting of the external surface requires low contact angle θ and low surface tension γ_{LV}. Wetting of the internal surface (i.e., penetration through pores) requires low θ but high γ_{LV}. These two conditions are incompatible and a compromise has to be made: $\gamma_{SV} - \gamma_{SL}$ must be kept at a maximum. γ_{LV} should be kept as low as possible but not too low.

The above conclusions illustrate the problem of choosing the best dispersing agent for a particular powder. This requires measurement of the above parameters as well as testing the efficiency of the dispersion process.

3.5.1
Rate of Penetration of Liquids: The Rideal–Washburn Equation

For horizontal capillaries (gravity neglected), the depth of penetration l in time t is given by the Rideal–Washburn equation [16, 17],

$$l = \left[\frac{rt\gamma_{LV}\cos\theta}{2\eta}\right]^{1/2} \tag{3.28}$$

To enhance the rate of penetration, γ_{LV} has to be made as high as possible, θ as low as possible, and η as low as possible. For dispersion of powders into liquids one should use surfactants that lower θ while not reducing γ_{LV} too much. The viscosity of the liquid should also be kept at a minimum. Thickening agents (such as polymers) should not be added during the dispersion process. It is also necessary to avoid foam formation during the dispersion process.

For a packed bed of particles, r may be replaced by K, which contains the effective radius of the bed and a turtuosity factor, which takes into account the complex path formed by the channels between the particles, that is,

$$l^2 = \left(\frac{kt\gamma_{LV}\cos\theta}{2\eta}\right)t \tag{3.29}$$

Thus a plot of l^2 versus t gives a straight line and from the slope of the line one can obtain θ. This is illustrated in Figure 3.12.

The Rideal–Washburn equation can be applied to obtain the contact angle of liquids (and surfactant solutions) in powder beds. K should first be obtained using a liquid that produces zero contact angle. This is discussed below.

3.5.2
Measurement of Contact Angles of Liquids and Surfactant Solutions on Powders

A packed bed of powder is prepared say in a tube fitted with a sintered glass at the end (to retain the powder particles). It is essential to pack the powder uniformly in the tube (a plunger may be used in this case). The tube containing the bed is immersed in a liquid that gives spontaneous wetting (e.g., a lower alkane), that is, the liquid gives a zero contact angle and $\cos\theta = 1$. By measuring the rate of pen-

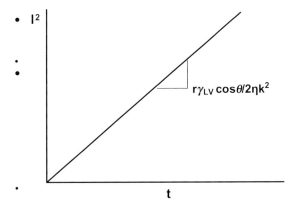

Figure 3.12 Variation of l^2 with t.

etration of the liquid (this can be carried out gravimetrically using for example a microbalance or a Kruss instrument) one can obtain K. The tube is then removed from the lower alkane liquid and left to stand for evaporation of the liquid. It is then immersed in the liquid in question and the rate of penetration is measured again as a function of time. Using Eq. (3.27), one can calculate $\cos\theta$ and hence θ.

3.5.3
Assessment of Wettability of Powders

3.5.3.1 Sinking Time, Submersion, or Immersion Test

This is by far the most simple (but qualitative) method for assessment of wettability of a powder by a surfactant solution. The time for which a powder floats on the surface of a liquid before sinking into the liquid is measured. 100 ml of the surfactant solution is placed in a 250 ml beaker (of internal diameter of 6.5 cm) and after 30 min. Standing 0.30 g of loose powder (previously screened through a 200-mesh sieve) is distributed with a spoon onto the surface of the solution. The time t for the 1- to 2-mm-thin powder layer to completely disappear from the surface is measured using a stop watch. Surfactant solutions with different concentrations are used and t is plotted versus surfactant concentration as is illustrated in Figure 3.13.

3.5.3.2 List of Wetting Agents for Hydrophobic Solids in Water

The most effective wetting agent is the one that gives a zero contact angle at the lowest concentration. For $\theta = 0_0$ or $\cos\theta = 1$, γ_{SL} and γ_{LV} have to be as low as possible. This requires a quick reduction of γ_{SL} and γ_{LV} under dynamic conditions during powder dispersion (this reduction should normally be achieved in less than 20 s). This requires fast adsorption of the surfactant molecules both at the LV and SL interfaces. It should be mentioned that the reduction of γ_{LV} is not always accompanied by simultaneous reduction of γ_{SL} and hence it is necessary to have information on both interfacial tensions which means that measurement of the contact angle is essential in selection of wetting agents. Measurement of γ_{SL} and γ_{LV} should be carried out under dynamic conditions (i.e., at very short times). In the absence

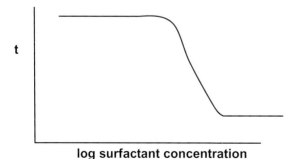

Figure 3.13 Sinking time as a function of surfactant concentration.

of such measurements, the sinking time described above could be applied as a guide for wetting agent selection. The most commonly used wetting agents for hydrophobic solids are listed below.

To achieve rapid adsorption the wetting agent should be either a branched chain with central hydrophilic group or a short hydrophobic chain with hydrophilic end group. The most commonly used wetting agents are the following:

Aerosol OT (diethylhexyl sulphosuccinate)

$$\begin{array}{c} C_2H_5 \quad\quad O \\ | \quad\quad\quad \| \\ C_4H_9CHCH_2-O-C-CH-SO_3Na \\ | \\ C_4H_9CHCH_2-O-C-CH_2 \\ | \quad\quad\quad \| \\ C_2H_5 \quad\quad O \end{array}$$

The above molecule has a low cmc of $0.7\,g\,dm^{-3}$ and at and above the cmc the water surface tension is reduced to $\sim25\,mN\,m^{-1}$ in less than 15 s.

An alternative anionic wetting agent is sodium dodecylbenzene sulfonate with a branched alkyl chain

$$\begin{array}{c} C_6H_{13} \\ | \\ CH_3-C-\bigcirc-SO_3Na \\ | \\ C_4H_9 \end{array}$$

The above molecule has a higher cmc ($1\,g\,dm^{-3}$) than Aerosol OT. It is also not very effective in lowering the surface tension of water reaching a value of $30\,mN\,m^{-1}$ at and above the cmc. It is, therefore, not as effective as Aerosol OT for powder wetting.

Several nonionic surfactants such as the alcohol ethoxylates can also be used as wetting agents. These molecules consist of a short hydrophobic chain (mostly C10) which is also branched. A medium chain polyethylene oxide (PEO) mostly consisting of six EO units or lower is used. These molecules also reduce the dynamic surface tension within a short time (<20 s) and they have a reasonably low cmc.

In all cases one should use the minimum amount of wetting agent to avoid interference with the dispersant that needs to be added to maintain the colloid stability during dispersion and on storage.

References

1 Zisman, W.A. (1964) *Adv. Chem. Ser.*, **43**, 1.
2 Good, R.J., and Girifalco, L.A. (1960) *J. Phys. Chem.*, **64**, 561.
3 Good, R.J. (1964) *Adv. Chem. Ser.*, **43**, 74.
4 Fowkes, F.M. (1964) *Adv. Chem. Ser.*, **43**, 99.
5 Gibbs, J.W. (1928) *The Collected Work of J. Willard Gibbs*, vol. 1, Longman, Harlow.

6 Smolders, C.A. (1960) *Rec. Trav. Chim.*, **80**, 650.
7 Dukhin, S.S., Kretzscmar, G., and Miller, R. (1995) *Dynamics of Adsorption at Liquid Interfaces*, Elsevier Publishers, Amsterdam.
8 Ward, A.F.H., and Tordai, L. (1946) *J. Phys. Chem.*, **14**, 453.
9 Panaitov, I., and Petrov, J.G. (1968/69). *Ann. Univ. Sofia, Fac. Chem.*, **64**, 385.
10 Hansen, R.S. (1960) *J. Phys. Chem.*, **64**, 637.
11 Miller, R., and Lunkenheimer, K. (1978) *Z. Phys. Chem.*, **259**, 863.
12 Zana, R. (1974) *Chem. Biol. Appl. Relaxation Spectrosc., Proc. NATO Adv. Study Inst., Ser. C*, **18**, 133.
13 Miller, R., Hoffmann, A., Hartmann, R., Schano, K.H., and Halbig, A. (1992) *Adv. Mater.*, **4**, 370.
14 Davies, J.T., and Rideal, E.K. (1969) *Interfacial Phenomena*, Academic Press, New York.
15 Kloubek, J., Friml, K., and Krejci, F. (1976) *Check. Chem. Commun.*, **41**, 1845.
16 Rideal, E.K. (1922) *Phil. Mag.*, **44**, 1152.
17 Washburn, E.D. (1921) *Phys. Rev.*, **17**, 273.

4
Structure of the Solid–Liquid Interface and Electrostatic Stabilization

4.1
Structure of the Solid–Liquid Interface

4.1.1
Origin of Charge on Surfaces

A great variety of processes occur to produce a surface charge.

4.1.1.1 Surface Ions

These are ions that have such a high affinity to the surface of the particles that they may be taken as part of the surface, for example, Ag^+ and I^- for AgI. For AgI in a solution of KNO_3, the surface charge σ_0 is given by the following expression:

$$\sigma_p = F(\Gamma_{Ag^+} - \Gamma_{I^-}) = F\Gamma_{AgNO_3} - \Gamma_{KI} \tag{4.1}$$

where F is the Faraday constant (96 500 C mol^{-1}) and Γ is the surface excess of ions (mol m^{-2}).

Similarly for an oxide such as silica or alumina in KNO_3, H^+ and OH^- may be taken as a part of the surface:

$$\sigma_0 = F(\Gamma_{H^+} - \Gamma_{OH^-}) = F(\Gamma_{HCl} - \Gamma_{KOH}) \tag{4.2}$$

The ions that determine the charge on the surface are termed potential determining ions.

Consider an oxide surface (Figure 4.1).

The charge depends on the pH of the solution. Below a certain pH, the surface is positive and above a certain pH the surface is negative. At a specific pH ($\Gamma_H = \Gamma_{OH}$) surface is uncharged; this is referred to as the point of zero charge (pzc).

The pzc depends on the type of the oxide. For an acidic oxide such as silica, the pzc is ~ pH 2–3. For a basic oxide such as alumina, pzc is ~ pH 9. For an amphoteric oxide such as titania, the pzc ~ pH 6. Some typical values of pzc for various oxides are given in Table 4.1.

In some cases, specifically adsorbed ions (that have nonelectrostatic affinity to the surface) "enrich" the surface but may not be considered as part of the surface,

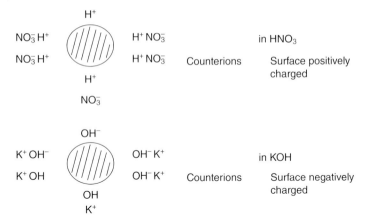

Figure 4.1 Schematic representation of oxide surface.

Table 4.1 Pzc values for some oxides.

Oxide	pzc
SiO_2 (precipitated)	2–3
SiO_2 (quartz)	3.7
SnO_2 (cassiterite)	5–6
TiO_2 (anatase)	6.2
TiO_2 (rutile)	5.7–5.8
RuO_2	5.7
α-Fe_2O_3 (hematite)	8.5–9.5
α-$FeO \cdot OH$ (goethite)	8.4–9.4
ZnO	8.5–9.5
γ-$Al(OH)_3$ (gibbsite)	8–9

for example, bivalent cations on oxides, cationic, and anionic surfactants on most surfaces [1].

4.1.1.2 Isomorphic Substitution

For example, with sodium montmorillonite, that is, replacement of cations inside the crystal structure by cations of lower valency, for example, Si^{4+} replaced by Al^{3+}. The deficit of one positive charge gives one negative charge. The surface of Na montmorillonite is negatively charged with Na^+ as counterions. This is schematically illustrated in Figure 4.2.

The surface charge + counterions from the electrical double layer.

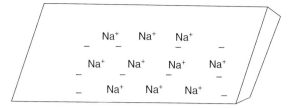

Figure 4.2 Schematic representation of clay particle.

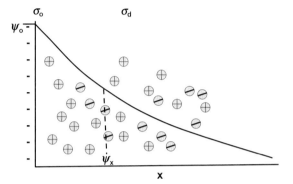

Figure 4.3 Schematic representation of the diffuse double layer according to Gouy and Chapman.

4.2
Structure of the Electrical Double Layer

4.2.1
Diffuse Double Layer (Gouy and Chapman)

The surface charge σ_o is compensated by unequal distribution of counterions (opposite in charge to the surface) and co-ions (same sign as the surface) that extend to some distance from the surface [2, 3]. This is schematically represented in Figure 4.3. The potential decays exponentially with the distance x. At low potentials

$$\psi = \psi_0 \exp-(\kappa x) \tag{4.3}$$

Note that when $x = 1/\kappa$, $\psi_x = \psi_0/e - 1/\kappa$ is referred to as the "thickness" of the "double layer."

The double-layer extension depends on electrolyte concentration and valency of the counterions:

$$n = \frac{n_0}{1+kn_0 t} \tag{4.4}$$

ε_r is the permittivity (dielectric constant); 78.6 for water at 25 °C. ε_0 is the permittivity of free space. k is the Boltzmann constant and T is the absolute temperature. n_0 is the number of ions per unit volume of each type present in bulk solution and Z_i is the valency of the ions. e is the electronic charge.

For 1:1 electrolyte (e.g., KCl)

C (mol dm^{-3})	10^{-5}	10^{-4}	10^{-3}	10^{-2}	10^{-1}
(1/κ) (nm)	100	33	10	3.3	1

The double-layer extension increases with decrease in electrolyte concentration.

4.2.2
Stern–Grahame Model of the Double Layer

Stern [4] introduced the concept of the nondiffuse part of the double layer for specifically adsorbed ions, the rest being diffuse in nature. This is schematically illustrated in Figure 4.4.

The potential drops linearly in the Stern region and then exponentially. Grahame distinguished two types of ions in the Stern plane, physically adsorbed counterions (outer Helmholtz plane), and chemically adsorbed ions (that loose part of their hydration shell) (inner Helmholtz plane).

4.3
Distinction between Specific and Nonspecific Adsorbed Ions

For the specifically adsorbed ions, the range of interaction is short, that is, these ions must reside at the distance of closest approach, possibly within the hydration shell. For the indifferent ions the situation is different and these ions are subjected to an attractive (for the counterions) or repulsive (for the co-ions) potential (energy = $\pm ZF\psi(x)/RT$). The space charge density due to these ions is high near the surface and decrease gradually with distance to its bulk value. Such a layer is

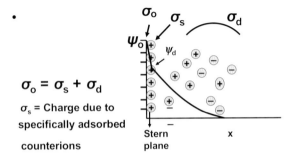

Figure 4.4 Schematic representation of the double layer according to Stern and Grahame.

the diffuse double layer described by Gouy and Chapman [2, 3]. Generally speaking a double layer contains a part that is specifically adsorbed and a diffuse part. Because of the finite size of the counterions there is always a charge-free layer near the surface.

4.4
Electrical Double-Layer Repulsion

When charged colloidal particles in a dispersion approach each other such that the double layers begin to overlap (particle separation becomes less than twice the double-layer extension), repulsion occurs. The individual double layers can no longer develop unrestrictedly, since the limited space does not allow complete potential decay [5].

This is illustrated in Figure 4.5 for two flat plates. The potential $\psi_{H/2}$ half way between the plates is no longer zero (as would be the case for isolated particles at $x \to \infty$). The potential distribution at an interparticle distance H is schematically depicted by the full line in Figure 4.5. The stern potential ψ_d is considered to be independent of the particle distance. The dashed curves show the potential as a function of the distance x to the Helmoltz plane, had the particles been at infinite distance.

For two spherical particles of radius R and surface potential ψ_0 and condition $\kappa R < 3$, the expression for the electrical double-layer repulsive interaction is given by [6]:

$$G_{el} = \frac{4\pi \varepsilon_r \varepsilon_0 R^2 \psi_0^2 \exp-(\kappa h)}{2R+h} \qquad (4.5)$$

where h is the closest distance of separation between the surfaces.

The above expression shows the exponential decay of G_{el} with h. The higher the value of κ (i.e., the higher the electrolyte concentration), the steeper the decay, as schematically shown in Figure 4.6.

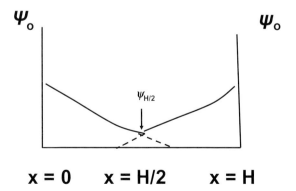

Figure 4.5 Schematic representation of double-layer interaction for two flat plates.

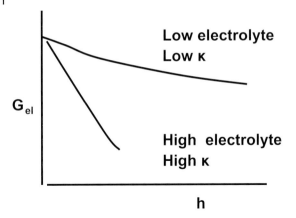

Figure 4.6 Variation of G_{el} with h at different electrolyte concentrations.

This means that at any given distance h, the double-layer repulsion decreases with increase of electrolyte concentration.

An important aspect of the double-layer repulsion is the situation during the particle approach. If at any stage the assumption is made that the double layers adjust to new conditions, so that equilibrium is always maintained then the interaction takes place at constant potential. This would be the case if the relaxation time of the surface charge is much shorter than the time the particles are in each other's interaction sphere as a result of Brownian motion. However, if the relaxation time of the surface charge is appreciably longer than the time particles are in each other's interaction sphere, the charge rather than the potential will be the constant parameter. The constant charge leads to larger repulsion than the constant potential case.

4.5
van der Waals Attraction

As is well known, atoms or molecules always attract each other at short distances of separation. The attractive forces are of three different types: dipole–dipole interaction (Keesom), dipole–induced-dipole interaction (Debye), and London dispersion force. The London dispersion force is the most important, since it occurs for polar and nonpolar molecules. It arises from fluctuations in the electron density distribution.

At small distances of separation r in vacuum, the attractive energy between two atoms or molecules is given by

$$G_{aa} = -\frac{\beta_{11}}{r^6} \tag{4.6}$$

β_{11} is the London dispersion constant.

4.5 van der Waals Attraction

For colloidal particles that are made of atom or molecular assemblies, the attractive energies have to be compounded. In this process, only the London interactions have to be considered, since large assemblies have neither a net dipole moment nor a net polarization. The result relies on the assumption that the interaction energies between all molecules in one particle with all the other are simply additive [7]. The interaction between two identical half-infinite plates at a distance H in vacuum is given by

$$G_A = -\frac{A_{11}}{12\pi h^2} \quad (4.7)$$

Whereas for two spheres in vacuum the result is

$$G_A = -\frac{A_{11}}{6}\left(\frac{2}{s^2-4} + \frac{2}{s^2} + \ln\frac{s^2-4}{s^2}\right) \quad (4.8)$$

A_{11} is known as the Hamaker constant and is defined by

$$A_{11} = \pi^2 q_{11}^2 \beta_{ii} \quad (4.9)$$

q_{11} is number of atoms or molecules of type-1 per unit volume, and $s = (2R+h)/R$. Equation (4.9) shows that A_{11} has the dimension of energy.

For very short distances ($h \ll R$), Eq. (4.8) may be approximated by

$$G_A = -\frac{A_{11}R}{12h} \quad (4.10)$$

When the particles are dispersed in a liquid medium, the van der Waals attraction has to be modified to take into account the medium effect. When two particles are brought from infinite distance to h in a medium, an equivalent amount of medium has to be transported the other way round. Hamaker forces in a medium are excess forces.

Consider two identical spheres 1 at a large distance apart in a medium 2 as is illustrated in Figure 4.7a. In this case, the attractive energy is zero. Figure 4.7b gives the same situation with arrows indicating the exchange of 1 against 2. Figure 4.7c shows the complete exchange that now shows the attraction between the two particles 1 and 1 and equivalent volumes of the medium 2 and 2.

The effective Hamaker constant for two identical particles 1 and 1 in a medium 2 is given by

$$A_{11(2)} = A_{11} + A_{22} - 2A_{12} = (A_{11}^{1/2} - A_{22}^{1/2})^2 \quad (4.11)$$

Equation (4.11) shows that two particles of the same material attract each other unless their Hamaker constant exactly matches each other. Equation (4.10) now becomes

$$G_A = -\frac{A_{11(2)}R}{12h} \quad (4.12)$$

where $A_{11(2)}$ is the effective Hamaker constant of two identical particles with Hamaker constant A_{11} in a medium with Hamaker constant A_{22}.

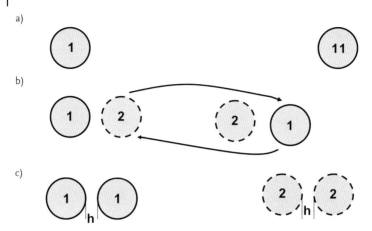

Figure 4.7 Schematic representation of interaction of two particles in a medium.

Table 4.2 Hamaker constant of some liquids.

Liquid	$A_{22} \times 10^{20}$ J
Water	3.7
Ethanol	4.2
Decane	4.8
Hexadecane	5.2
Cyclohexane	5.2

Table 4.3 Effective Hamaker constant $A_{11(2)}$ of some particles in water.

System	$A_{11(2)} \times 10^{20}$ J
Fused quartz/water	0.83
Al_2O_3/water	5.32
Copper/water	30.00
Poly(methylmethacrylate)/water	1.05
Poly(vinylchloride)/water	1.03
Poly(tetrafluoroethylene)/water	0.33

In most cases, the Hamaker constant of the particles is higher than that of the medium. Examples of Hamaker constant for some liquids are given in Table 4.2. Table 4.3 gives values of the effective Hamaker constant for some particles in some liquids. Generally speaking, the effect of the liquid medium is to reduce the Hamaker constant of the particles below its value in vacuum (air).

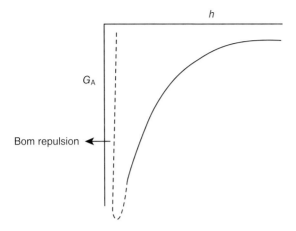

Figure 4.8 Variation of G_A with h.

G_A decreases with the increase of h as schematically shown in Figure 4.8.

As shown in Figure 4.8, V_A increases very sharply with h at small h values. The capture distance can be defined at which all the particles become strongly attracted to each other (coagulation). At very short distances, the Born repulsion appears.

The Hamaker approach, referred to as a "microscopic" theory is based on the interactions between pairs of atoms or molecules. The more accurate "macroscopic" approach originally suggested by Lifshits and described in detail by Mahanty and Ninham [8] is based on the principle that the spontaneous electromagnetic fluctuations in two particles become correlated when the latter approach each other, causing a decrease in the free energy of the system. The elaboration of this theory is rather complex and its application requires extensive data on the electromagnetic interaction energies. Nevertheless, the theory allows for the important conclusion that the most qualitative aspects of the "microscopic" theory given by Eqs. (4.7)–(4.12) are fully confirmed. The only exception concerns the decay of G_A with h at large separations. Owing to the time required for electromagnetic waves to cover the distance between the particles, the h^{-2} dependence in Eq. (4.7) gradually changes to h^{-3} dependence at large separations, a phenomenon known as retardation.

4.6
Total Energy of Interaction

4.6.1
Deryaguin–Landau–Verwey–Overbeek Theory [9–11]

Combination of G_{el} and G_A results in the well-known theory of stability of colloids (DLVO Theory) [8, 9]:

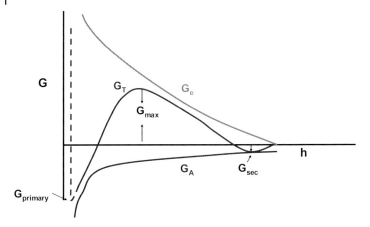

Figure 4.9 Schematic representation of the variation of G_T with h according to the DLVO theory.

$$G_T = G_{el} + G_A \qquad (4.13)$$

A plot of G_T versus h is shown in Figure 4.9, which represents the case at low electrolyte concentrations, that is, strong electrostatic repulsion between the particles. G_{el} decays exponentially with h, that is, $G_{el} \to 0$ as h becomes large. G_A is ∞ $1/h$, that is, G_A does not decay to 0 at large h.

At long distances of separation, $G_A > G_{el}$ resulting in a shallow minimum (secondary minimum). At very short distances, $G_A \gg G_{el}$ resulting in a deep primary minimum.

At intermediate distances, $G_{el} > G_A$ resulting in energy maximum, G_{max}, whose height depends on ψ_0 (or ψ_d) and the electrolyte concentration and valency.

At low electrolyte concentrations ($<10^{-2}$ mol dm^{-3} for a 1:1 electrolyte), G_{max} is high ($>25\,kT$) and this prevents particle aggregation into the primary minimum. The higher the electrolyte concentration (and the higher the valency of the ions), the lower the energy maximum.

Under some conditions (depending on the electrolyte concentration and particle size), flocculation into the secondary minimum may occur. This flocculation is weak and reversible. By increasing the electrolyte concentration, G_{max} decreases till at a given concentration it vanishes and particle coagulation occurs. This is illustrated in Figure 4.10 that shows the variation of G_T with h at various electrolyte concentrations.

Since approximate formulae are available for G_{el} and G_A, quantitative expressions for $G_T(h)$ can also be formulated. These can be used to derive expressions for the coagulation concentration, which is, that concentration that causes every encounter between two colloidal particles to lead to destabilization. Verwey and Overbeek [10] introduced the following criteria for transition between stability and instability:

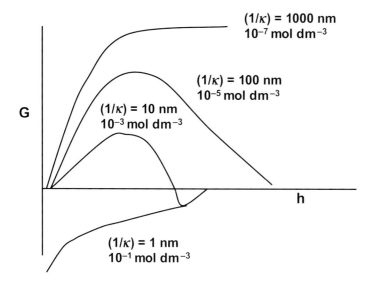

Figure 4.10 Variation of G with h at various electrolyte concentrations.

$$G_T (= G_{el} + G_A) = 0 \tag{4.14}$$

$$\frac{dG_T}{dh} = 0 \tag{4.15}$$

$$\frac{dG_{el}}{dh} = -\frac{dG_A}{dh} \tag{4.16}$$

Using the equations for G_{el} and G_A, the critical coagulation concentration (c.c.c.) could be calculated as will be shown below. The theory predicts that c.c.c. is directly proportional to the surface potential ψ_0 and inversely proportional to the Hamaker constant A and the electrolyte valency Z. As will be shown below, the c.c.c is inversely proportional to Z^6 at high surface potential and inversely proportional to Z^6 at low surface potential.

4.7
Flocculation of Suspensions

As discussed before, the condition for kinetic stability is $G_{max} > 25\,kT$. When $G_{max} < 5\,kT$, flocculation occurs. Two types of flocculation kinetics may be distinguished: fast flocculation with no energy barrier and slow flocculation when an energy barrier exists.

The fast flocculation kinetics was treated by Smoluchowski [12], who considered the process to be represented by second-order kinetics and the process is simply diffusion controlled. The number of particles n at any time t may be related to the initial number (at $t = 0$) n_0 by the following expression:

$$n = \frac{n_0}{1 + kn_0 t} \tag{4.17}$$

where k is the rate constant for fast flocculation that is related to the diffusion coefficient of the particles D, that is,

$$k = 8\pi DR \tag{4.18}$$

D is given by the Stokes–Einstein equation:

$$D = \frac{kT}{6\pi\eta R} \tag{4.19}$$

Combining Eqs. (4.12) and (4.13),

$$k = \frac{4}{3}\frac{kT}{\eta} = 5.5 \times 10^{-18} \; m^3 s^{-1} \; \text{for water at } 25°C \tag{4.20}$$

The half-life $t_{1/2}$ ($n = (1/2) n_0$) can be calculated at various n_0 or volume fraction φ as give in Table 4.4.

The slow flocculation kinetics was treated by Fuchs [13] who related the rate constant k to the Smoluchowski rate by the stability constant W:

$$W = \frac{k_0}{k} \tag{4.21}$$

W is related to G_{max} by the following expression:

$$W = \frac{1}{2}k\exp\left(\frac{G_{max}}{kT}\right) \tag{4.22}$$

Since G_{max} is determined by the salt concentration C and valency, one can derive an expression relating W to C and Z [14],

$$\log W = -2.06 \;\; 10^9 \left(\frac{R\gamma^2}{Z^2}\right)\log C \tag{4.23}$$

Table 4.4 Half-life of suspension flocculation.

R (μm)	φ			
	10^{-5}	10^{-2}	10^{-1}	5×10^{-1}
0.1	765 s	76 ms	7.6 ms	1.5 ms
1.0	21 h	76 s	7.6 s	1.5 s
10.0	4 month	21 h	2 h	25 m

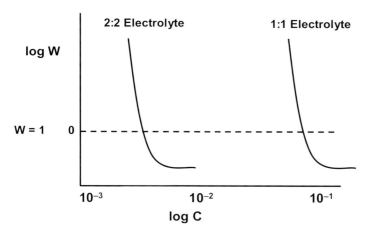

Figure 4.11 Log W–log C curves for electrostatically stabilized emulsions.

where γ is a function that is determined by the surface potential ψ_0,

$$\gamma = \left[\frac{\exp(Ze\psi_0/kT)-1}{\exp(Ze\psi_0/kT)+1} \right] \tag{4.24}$$

Plots of log W versus log C are shown in Figure 4.11. The condition log $W = 0$ ($W = 1$) is the onset of fast flocculation. The electrolyte concentration at this point defines the critical flocculation concentration c.c.c. Above the c.c.c, $W < 1$ (due to the contribution of van der Waals attraction that accelerates the rate above the Smoluchowski value). Below the c.c.c, $W > 1$ and it increases with decrease of electrolyte concentration. The above figure also shows that the c.c.c. decreases with increase of valency. At low surface potentials, c.c.c. $\propto 1/Z^2$. This is referred to as the Schultze–Hardy rule.

Another mechanism of flocculation is that involving the secondary minimum (G_{min}) which is few kT units – in this case flocculation is weak and reversible and hence one must consider both the rate of flocculation (forward rate k_f) and deflocculation (backward rate k_b). In this case, the rate or decrease of particle number with time is given by the expression

$$-\frac{dn}{dt} = -k_f n^2 + k_b n \tag{4.25}$$

The backward reaction (break-up of weak flocs) reduces the overall rate of flocculation.

Another process of flocculation that occurs under the shearing conditions is referred to as orthokinetic (to distinguish it from the diffusion controlled perikinetic process). In this case, the rate of flocculation is related to the shear rate by the expression

$$-\frac{dn}{dt} = \frac{16}{3}\alpha^2 \gamma R^3 \tag{4.26}$$

where α is the collision frequency, that is, the fraction of collisions that result in permanent aggregates.

4.8
Criteria for Stabilization of Dispersions with Double-Layer Interaction

The two main criteria for stabilization are: (i) high surface or stern potential (zeta potential), high surface charge. As shown in Eq. (4.5), the repulsive energy G_{el} is proportional to ψ_0^2. In practice, ψ_0 cannot be directly measured and, therefore, one instead uses the measurable zeta potential as will be discussed in Chapter 5. (ii) Low electrolyte concentration and low valency of counter and co-ions. As shown in Figure 4.10, the energy maximum increases with decrease of electrolyte concentration. The latter should be lower than $10^{-2}\,mol\,dm^{-3}$ for 1:1 electrolyte and lower than $10^{-3}\,mol\,dm^{-3}$ for 2:2 electrolyte. One should ensure that an energy maximum in excess of $25\,kT$ should exist in the energy–distance curve. When $G_{max} \gg kT$, the particles in the dispersion cannot overcome the energy barrier, thus preventing coagulation. In some cases, particularly with large and asymmetric particles, flocculation into the secondary minimum may occur. This flocculation is usually weak and reversible and may be advantageous for preventing the formation of hard sediments.

References

1 Lyklema, J. (1987) Structure of the solid/liquid interface and the electrical double layer, in *Solid/Liquid Dispersions* (ed. Th.F. Tadros), Academic Press, London, pp. 63–90.
2 Gouy, G. (1910) *J. Phys.*, **9**, 457; (1917) *Ann. Phys.*, **7**, 129.
3 Chapman, D.L. (1913) *Philos. Mag.*, **25**, 475.
4 Stern, O. (1924) *Z. Electrochem.*, **30**, 508.
5 Grahame, D.C. (1947) *Chem. Rev.*, **41**, 44.
6 Bijesterbosch, B.H. (1987) Stability of solid–liquid dispersions, in *Solid/Liquid Dispersions* (ed. Th.F. Tadros), Academic Press, London, pp. 91–109.
7 Hamaker, H.C. (1937) *Physica*, **4**, 1058.
8 Mahanty, J., and Ninham, B.W. (1976) *Dispersion Forces*, Academic Press, London.
9 Deryaguin, B.V., and Landau, L. (1941) *Acta Physicochem. USSR*, **14**, 633.
10 Verwey, E.J.W., and Overbeek, J.Th.G. (1948) *Theory of Stability of Lyophobic Colloids*, Elsevier, Amsterdam.
11 Kruyt, H.R. (ed.) (1952) *Colloid Science*, vol. I, Elsevier, Amsterdam.
12 Smoluchowski, M.V. (1927) *Z. Phys. Chem.*, **92**, 129.
13 Fuchs, N. (1936) *Z. Phys.*, **89**, 736.
14 Reerink, H., and Overbeek, J.Th.G. (1954) *Discuss. Faraday Soc.*, **18**, 74.

5
Electrokinetic Phenomena and Zeta Potential

As mentioned in Chapter 4, one of the main criteria for electrostatic stability is the high surface or zeta potential that can be experimentally measured as will be discussed below. Before describing the experimental techniques for measuring the zeta potential it is essential to consider the electrokinetic effects in some detail, describing the theories that can be used to calculate the zeta potential from the particle electrophoretic mobility [1].

Electrokinetic effects are the direct result of charge separation at the interface between two phases. This is illustrated in Figure 5.1.

Consider a negatively charged surface; positive ions (counterions) are attracted to the surface, whereas negative ions (co-ions) are repelled. This is schematically shown in Figure 5.2. If the surface of phase I is negatively charged, its electrostatic potential will be negative relative to the bulk of phase II. If phase II is a liquid containing dissolved ions, then as one moves into phase II, the potential will decrease more or less regularly, until it becomes constant in bulk liquid far from the surface of phase I. This constant potential is usually referred to as the zeta potential. The constant potential is usually reached at a distance in the region of 5–200 nm, depending on electrolyte concentration. In most colloid systems the point p shown in Figure 5.2 is at a distance about 1–50 nm from the surface.

The accumulation of excess positive ions causes a gradual reduction in the potential from its value ψ_0 at the surface to 0 in the bulk solution. At point p from the surface, one can define a potential ψ_x. As we will see later, the zeta potential is taken at the point of the "shear plane" (that is an imaginary plane from the particle surface at which one of the phases move tangentially past the second phase). The region where the liquid has a negative electrostatic potential will accumulate an excess of positive ions (counterions) and repel negative ions of the electrolyte (co-ions). It is this excess positive ions that gradually lowers the electrostatic potential (and the electric field) to zero in the bulk solution. The arrangement of negative charges on the surface of phase I and the charges in phase II (counterions and co-ions) is referred to as the electrical double layer at the interface.

Electrokinetic effects arise when one of the two phases is caused to move tangentially past the second phase. The tangential motion can be caused by electric field, forcing a liquid in a capillary, forcing a liquid in a plug of particles or by

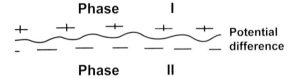

Figure 5.1 Schematic representation of charge separation.

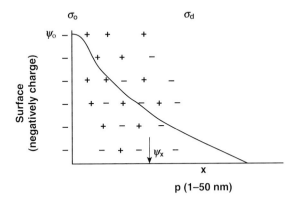

Figure 5.2 Schematic representation of charge accumulation at an interface.

gravitational field on the particles. This leads to four different types of electrokinetic phenomena:

i) **Electrophoresis:** In electrophoresis, the movement of one phase is induced by the application of an electric field (with a field strength E/l (V m^{-1})), where E is the potential difference applied and l is the distance between the two electrodes. This is schematically illustrated in Figure 5.3.

In electrophoresis, a fluid moves with respect to a solid, liquid, or gas surface, that is, movement of one phase induced by application of an external electric field. One measures the particle velocity v (m s^{-1}) of the particles, droplets or air bubbles from which the electrophoretic mobility u (velocity divided by field strength) can be calculated:

$$u = \frac{v}{(E/l)} \text{ m}^2\text{V}^{-1}\text{s}^{-1} \tag{5.1}$$

where E is the applied potential and l is the distance between the two electrodes; E/l is the field strength. For example, if the surface charge on the capillary surface is negative, the counterions that are positive will move toward the cathode.

ii) **Electroosmosis:** In this case, the solid is kept stationary (e.g., a capillary or porous plug), whereas the liquid is allowed to move under the influence of an electric field. The electric field acts on the charges (ions) in the liquid. When these move, they drag liquid with them; one observes movement of the liquid along the capillary from one electrode to the other. For example, if the surface charge on the capillary surface is negative, the counterions that are positive will move toward the cathode.

iii) **Streaming potential:** The liquid is forced through a capillary or a porous plug (containing the particles) under the influence of a pressure gradient. The excess charges near the wall (or the surface of particles in the plug) are carried along by the liquid flow, thus producing an electric field that can be measured by using electrodes and an electrometer.

iv) **Sedimentation potential (Dorn effect):** Particles (in a suspension) or droplets (in an emulsion) or gas bubbles in foam are allowed to settle or rise under the influence of gravity or the centrifugal field. When the particles move (up or down depending on the density difference between the particles and medium), they leave behind their ionic atmosphere. A potential difference (sedimentation potential) develops in the direction of motion. If two electrodes are place in the sedimentation tube, one can measure a potential difference as a result of charge separation.

In this chapter, only electrophoresis will be discussed since this is the most commonly used method for dispersions, allowing one to measure the particle mobility that can be converted to the zeta potential using theoretical treatments.

In all electrokinetic phenomena [1], a fluid moves with respect to a solid surface. One needs to derive a relationship between fluid velocity (which varies with distance from the solid) and the electric field in the interfacial region. The main problem in any analysis of electrokinetic phenomena is defining the plane at which the liquid begins to move past the surface of the particle, droplet, or air bubble. This is defined as the "shear plane," which at some distance from the surface. One usually defines an "imaginary" surface close to the particle surface within which the fluid is stationary. The point just outside this imaginary surface is described as surface of shear and the potential at this point is described as the zeta potential (ζ). A schematic representation of the surface of shear, the surface and zeta potential is shown in Figure 5.3.

The exact position of the plane of shear is not known; it is usually in the region of few A. In some cases one may equate the shear plane with the Stern plane (the center of specifically adsorbed ions) although this may be an underestimate of its location. Several layers of liquid may be immobilized at the particle surface (which means that the shear plane is farther apart from the Stern plane). The particle, droplet, or air bubble plus its immobile liquid layer forms the kinetic unit that moves under the influence of the electric field. The viscosity of the liquid in the immobile sheath around the particles (η') is much larger than the bulk viscosity η. The permittivity of the liquid in this liquid sheath ε' is also lower than the bulk permittivity (due to dielectric saturation in this layer). In the absence of specific adsorption, the assumption is usually made that $\zeta \sim \psi_0$. The latter potential is the value that is commonly used to calculate the repulsive energy between two particles.

It is important to understand the relationship between the zeta potential and the potential distribution across the interface. For that purpose it is useful to

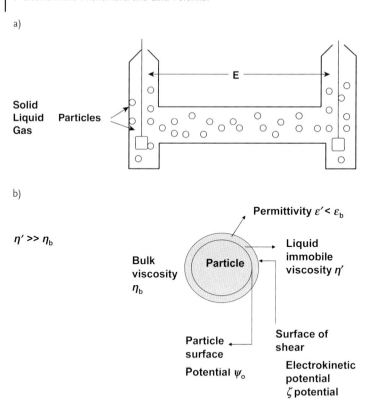

Figure 5.3 Schematic setup for electrophoresis.

consider the structure of the electrical double layer. As discussed in Chapter 4, one of the earliest pictures of the double layer is that due to Gouy and Chapman [2, 3], usually referred to as the diffuse double layer concept. In this picture, it was assumed that the charge on the surface is "smeared out" and this charge is compensated by a diffuse layer of counter- and co-ions that extends to a distance from the surface (that depends on electrolyte concentration and valency). A schematic representation of the diffuse double layer for a flat negatively charged surface was shown in Figure 5.2. In this case, the negative surface charge is compensated in the bulk solution by unequal distribution of counter- and co-ions (excess counterions and deficit of co-ions).

The surface potential decays exponentially with the distance x from the particle surface. For low potentials ($\psi_0 < 25$ mV), the potential ψ is related to the surface potential ψ_0 by the simple expression:

$$\psi = \psi_0 \exp(-\kappa x) \tag{5.2}$$

where κ is the Debye–Huckel parameter that is related to electrolyte concentration and valency:

Table 5.1 Double layer thickness for various concentrations of 1:1 electrolyte.

C_i (mol dm^{-3})	$(1/\kappa)$ (nm)
10^{-5}	100
10^{-4}	33
10^{-3}	10
10^{-2}	3.3
10^{-1}	1.0

$$\kappa^2 = \left(\frac{e^2 \sum n_i^2 z_i^2}{\varepsilon \varepsilon_0 kT}\right) \quad (5.3)$$

where e is the electronic charge n_i is the number of ions per unit volume, z_i is the valency of each type of ion, ε is the relative permittivity, ε_0 is the permittivity of free space, k is the Boltzmann constant, and T is the absolute temperature.

For water at 25 °C,

$$\kappa = 3.88 I^{1/2} (\text{nm}^{-1}) \quad (5.4)$$

where I is the ionic strength,

$$I = (1/2) \sum c_i z_i^2 \quad (5.5)$$

where c_i is the concentration of ion i in mol dm^{-3}. Note that the dimension of κ is in reciprocal length and $1/\kappa$ is referred to as the double layer thickness. It is clear from Eq. (5.2) that when $\kappa = 1/x$, $\psi = \psi_0/e$; $(1/\kappa)$ is referred to as the double layer extension or thickness that depends on c_i and z_i. As an illustration, Table 5.1 shows the values of $(1/\kappa)$ for various concentrations of 1:1 electrolyte (e.g., NaCl). It can be seen from Table 5.1 that as the electrolyte concentration increases, the thickness of the double layer $(1/\kappa)$ decreases. This amounts to compression of the double layer with increase in electrolyte concentration.

5.1
Stern–Grahame Model of the Double Layer

Stern [4] introduced the concept of the nondiffusive part of the double layer for specifically adsorbed ions, the rest being diffuse in nature. The potential drops linearly in the Stern region and then exponentially. Grahame [5] distinguished two types of ions in the Stern plane – physically adsorbed counterions, outer Helmholtz plane, and chemically adsorbed ions (that loose part of their hydration shell), inner Helmholtz plane.

The surface potential ψ_0 decays linearly with x till ψ_s (the position of the inner Helmholtz plane) and then ψ_d (the position of the outer Helmholtz plane) and then exponentially with further decrease in x. One usually equates ψ_d with the zeta potential ζ, although the exact value of ζ cannot be assigned from the above picture since the position of the shear plane is not identified in these double layer theories. Clearly if there is specific adsorption of counter- or co-ions in the IHP, the above equality is not justified.

Measurement of zeta potential (ζ) is valuable in determining the properties of dispersions. In addition it has many other applications in various fields: electrode kinetics, electrodialysis, corrosion, adsorption of surfactants and polymers, crystal growth, mineral flotation, and particle sedimentation.

Although measurement of particle mobility is fairly simple (particularly with the development of automated instruments), the interpretation of the results is not simple. The calculation of zeta potential from particle mobility is not straightforward since this depends on the particle size and shape as well as on the electrolyte concentration. For simplicity we will assume that the particles are spherical.

5.2
Calculation of Zeta Potential from Particle Mobility

5.2.1
von Smoluchowski (Classical) Treatment [6]

von Smoluchowski [6] considered the movement of the liquid adjacent to a flat, charged surface under the influence of an electric field parallel to the surface (i.e., electroosmotic flow of the liquid). If the surface is negatively charged, there will be a net excess of negative ions in the adjacent liquid and as they move under the influence of the applied field they will draw the liquid along with them. The surface of shear may be taken as a plane parallel to the surface and distant δ from it. The velocity of the liquid in the direction parallel to the wall, v_z, rises from a value of zero at the plane of shear to a maximum value, v_{eo}, at some distance from the wall, after which it remains constant. This is illustrated in Figure 5.4. v_{eo} is called the electroosmotic velocity of the liquid. The electrical potential ψ changes from its maximum negative value (ζ) at the shear plane to zero when v_z reaches v_{eo}.

Consider a volume element of area A and thickness dx, as shown in Figure 5.5. The applied electric force in the z-direction is $E_z Q$, where E_z is the field strength in the z-direction and Q is the charge density in the volume element that is equal to ρdAx, where ρ is the charge density per unit volume. This electric force is balanced by the hydrodynamic force at the liquid surfaces, that is,

$$E_z Q = E_z \rho A\, dx = \eta A \left(\frac{dv_x}{dx}\right)_x - \eta A \left(\frac{dv_x}{dx}\right)_{x+dx} \tag{5.6}$$

Equation (5.6) can be written as

$$E_z \rho\, dx = -\eta \left(\frac{d^2 v_x}{dx^2}\right) dx \tag{5.7}$$

5.2 Calculation of Zeta Potential from Particle Mobility

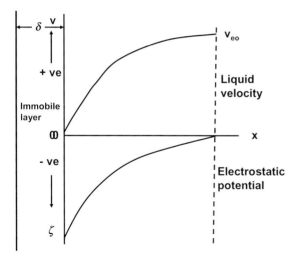

Figure 5.4 Distribution of electrostatic potential near a charged surface and the resulting elecroosmotic velocity under an applied field.

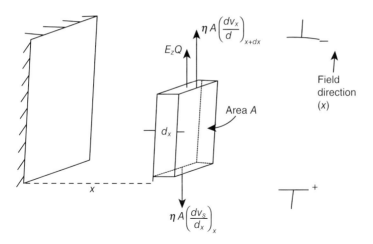

Figure 5.5 Force on a volume element of a liquid of area A containing charge Q.

From Poisson's equation,

$$\frac{d^2\psi}{dx^2} = -\frac{\rho}{\varepsilon\varepsilon_0} \tag{5.8}$$

Combining Eqs. (5.7) and (5.8),

$$E_z \varepsilon\varepsilon \frac{d^2\psi}{dx^2} dx = \eta \frac{d^2 v_x}{dx^2} dx \tag{5.9}$$

5 Electrokinetic Phenomena and Zeta Potential

Equation (5.9) can be integrated twice from a point from the solid where $\psi = 0$ and $v_z = v_{eo}$ up to the shear plane where $v_z = 0$ and $\psi = \zeta$, using the fact that for the first integration both $d\psi/dx$ and dv_x/dx are zero from the surface, that is,

$$\frac{v_{eo}}{E_z} = u_E = -\frac{\varepsilon\varepsilon_0 \zeta}{\eta} \tag{5.10}$$

Where u_E is the elecroosmotic mobility. The negative sign indicates that when ζ is negative the space charge is positive and the liquid flows toward the negative electrode.

In electroosmotic mobility experiments one usually measures the total volume of liquid transported say in a capillary on application of the electric field. For a capillary with constant cross section with radius r, the volume displaced per unit time, V, is given by

$$V = \pi r^2 v_{eo} = \frac{\pi r^2 \varepsilon\varepsilon_0 \zeta E_z}{\eta} \tag{5.11}$$

The electric current, i, transported by the liquid is given by Ohm's law:

$$\frac{i}{E_z} = \pi r^2 \lambda_0 \tag{5.12}$$

Here λ_0 is the electrical conductivity.

Combining Eqs (5.11) and (5.12),

$$\frac{V}{i} = \frac{\varepsilon\varepsilon_0 \zeta}{\eta \lambda_0} \tag{5.13}$$

In Eq. (5.13), the units can be in S.I., $\varepsilon_0 = 8.854 \times 10^{-12}\,CV^{-1}m^{-1}$, and if ζ is in volts, η in Pa s ($Nm^{-2}s$), V/i is obtained in $m^3 C^{-1}$ or $m^3 s^{-1}$ per ampere of current used.

One may also use mixed units to obtain ζ in mV:

$$\frac{V(cm^3 s^{-1})}{i(mA)} = \frac{8.854 \times 10^{-13} \varepsilon \zeta(mV)}{\eta(Poise) x \lambda_0(ohm^{-1}cm^{-1})} \tag{5.14}$$

In Eqs. (5.12)–(5.14), the assumption is made that the current is transported by the bulk liquid, that is, the contribution from the conductance near the wall or through the solid (surface conductance) is small. However, if the contribution from the surface conductance is significant the accumulated charge in the double layer may lead to an unusually high conductivity, especially at low electrolyte concentration. In this case, Eq. (5.12) has to be modified to take into account the surface conductance λ_s:

$$\frac{i}{E_z} = \pi r^2 \lambda_0 + 2\pi r \lambda_s \tag{5.15}$$

Note that λ_s is in ohm^{-1}.

Equation (5.13) then becomes

$$\frac{V}{i} = \frac{\varepsilon\varepsilon\zeta}{\eta(\lambda_0 + 2\lambda_s/r)} \quad (5.16)$$

The specific surface conductivity values are of the order of 10^{-9}–10^{-8} for water in glass capillaries so that significant effects on ζ-potential can be expected in 1 mm capillaries at electrolyte concentrations below about $10^{-3.5}$ mol dm^{-3}.

The above treatment can be applied to the electrophoretic motion of a large particle with a thin double layer ($\kappa R \gg 1$). The liquid is regarded as fixed so that the particle moves in the opposite direction from Eq. (5.10):

$$u_E = \frac{\varepsilon_r \varepsilon_0 \zeta}{\eta} \quad (5.17)$$

u_E is the electrophoretic mobility (Smoluchowski equation). For water at 25 °C, ε_r is the relative permittivity of the medium; 78.6, ε_0 is the permittivity of free space; 8.85×10^{-12} F m^{-1} and η is the viscosity of the medium, 8.9×10^{-4}:

$$\zeta = 1.282 \times 10^6 \, u \quad (5.18)$$

u is expressed in m^2 V^{-1} s^{-1} and ζ in volts.

Equation (5.17) applies to the case where the particle radius R is much larger than the double layer thickness ($1/\kappa$), that is, $\kappa R \gg 1$. This is generally the case for particles that are greater than 0.5 mµ (when the 1:1 electrolyte concentration is lower than 10^{-3} mol dm^{-3}, that is, $\kappa R > 10$),

5.2.2
The Huckel Equation [7]

Soon after the publication by Debye and Huckel of the theory of behavior of strong electrolytes, Huckel [7] reexamined the electrophoresis problem and obtained a significantly different result from Smoluchowski equation:

$$u = \frac{2}{3} \frac{\varepsilon_r \varepsilon_0 \zeta}{\eta} \quad (5.19)$$

The above equation applies for small particles (<100 nm) and thick double layers (low electrolyte concentration), that is, for the case $\kappa R < 1$.

Equation (5.19) can be simply derived by balancing the electric force on the particle, QE^z, with the frictional force given by Stokes' law ($6\pi \eta R v_E$), that is,

$$QE_z = 6\pi \eta R v_E \quad (5.20)$$

$$u_E = \frac{v_E}{E_z} = \frac{Q}{6\pi \eta R} \quad (5.21)$$

The electric charge Q is given in the following equation:

$$Q = 4\pi \varepsilon \varepsilon_0 (1 + \kappa R) \zeta \quad (5.22)$$

Combining Eqs. (5.21) and (5.22) one obtains

$$u_E = \frac{2\varepsilon\varepsilon_0 \zeta}{3\eta}(1+\kappa R) \qquad (5.23)$$

when $\kappa R \ll 1$, that is, small particles with relatively thick double layers, Eq. (5.23) reduces to Eq. (5.19).

A more rigorous derivation of Eq. (5.19) was given by Overbeek and Bijesterbosch [8]. The action of the electric field on the double layer, causing the liquid to move in accordance with Eq. (5.10) is called electrophoretic retardation because it causes a reduction in the velocity of the migrating particle. Smoluchowski's treatment [6] assumes that this is the dominant force and that the particle's motion is equal and opposite to the liquid motion. Huckel [7], on the other hand, also made proper allowance for the electrophoretic retardation in his analysis. However, as mentioned earlier, Eq. (5.19) is only valid for small values of κR when electrophoretic retardation is relatively unimportant and the main retarding force is the frictional resistance of the medium. The electrophoretic retardation at small κR remains important in the description of electrolyte conduction. In this case, one must consider the movement of ions of both positive and negative sign and the calculation of the interaction effects for large number of ions. In electrophoresis one considers only the particle that is regarded as isolated in an infinite medium. For large particles with thin double layers, essentially all of the elecrophoretic retardation is communicated directly to the particle.

5.2.3
Henry's Treatment [9]

Henry [9] accounted for the discrepancy between Scmoluchowski and Huckel's treatment by considering the electric field in the neighborhood of the particle. Huckel disregarded the deformation of the electric field by the particle, whereas Smoluchowski assumed the field to be uniform and everywhere parallel to the particle surface. As shown in Figure 5.6 these two assumptions are justified in the extreme cases of $\kappa R \ll 1$ and $\kappa R \ll 1$, respectively.

Henry [9] showed that when the external field is superimposed on the local field around the particle, the following expression for the mobility is used:

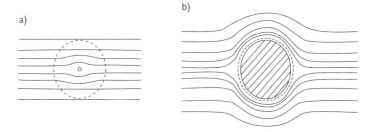

Figure 5.6 Effect of a nonconducting particle on the applied field. (a) $\kappa R \ll 1$; (b) $\kappa R \ll 1$. The broken line is at a distance $(1/\kappa)$ from the particle surface.

Table 5.2 Henry's correction factor $f(\kappa R)$.

κR	0	1	2	3	4	5	10	25	100	∞
$f(\kappa R)$	1.0	1.027	1.066	1.101	1.133	1.160	1.239	1.370	1.460	1.500

$$u = \frac{2}{3}\frac{\varepsilon_r \varepsilon_0 \zeta}{\eta} f(\kappa R) \tag{5.24}$$

The function $f(\kappa R)$ depends also on the particle shape. Values of $f(\kappa R)$ at increasing values of κR are given in Table 5.2.

Henry's calculations are based on the assumption that the external field can be superimposed on the field due to the particle and hence it can only be applied for low potentials ($\zeta < 25\,\text{mV}$). It also does not take into account the distortion of the field induced by the movement of the particle (relaxation effect).

Wiersema *et al.* [10] introduced two corrections for the Henry's treatment, namely the relaxation and retardation (movement of the liquid with the double layer ions) effects. (i) Distortion of the field induced by the movement of the particles (distortion of the double layer symmetry and its reformation). This is referred to as the relaxation effect. (ii) Movement of the liquid with the double layer ions, which results in reduction of the mobility of the integrating particles. This is referred to as the retardation effect. By considering these two effects, Wiersema *et al.* [10] derived exact expressions for the relationship between mobility and zeta potential for all κR values and any value of ζ-potential. Numerical tabulation of the relation between mobility and zeta potential has been given by Ottewill and Shaw [11]. Such tables are useful for conversion of u to ζ at all practical values of κR.

5.3
Measurement of Electrophoretic Mobility and Zeta Potential

5.3.1
Ultramicroscopic Technique (Microelectrophoresis)

This is the most commonly used method since it allows direct observation of the particles using an ultramicroscope (suitable for particles that are larger than 100 nm). Microelectrophoresis has many advantages since the particles can be measured in their normal environment. It is preferable to dilute the suspension with the supernatant liquid that can be produced by centrifugation. Basically, a dilute suspension is placed in a cell (that can be circular or rectangular) consisting of a thin walled (~100 μm) glass tube that is attached to two larger bore tubes with sockets for placing the electrodes. The cell is immersed in a thermostat bath (accurate to ±0.1 °C) that contains attachment for illumination and a microscope

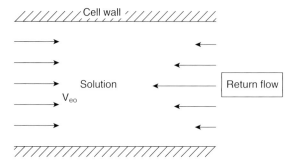

Figure 5.7 Flow conditions within a closed cylindrical electrophoretic cell with electric field applied.

objective for observing the particles. It is also possible to use a video camera for directly observing the particles.

Since the glass walls are charged (usually negative at practical pH measurements), the solution in the cell will in general experience electroosmotic flow. Thus, the observed motion of the particle when the field is applied, v_p, is the sum of its true velocity, v_E, and the total liquid velocity, v_l. The latter varies with the distance r from the axis in accordance with Poiseuille's equation:

$$v_l = p\frac{(a^2 - r^2)}{4\eta l} \tag{5.25}$$

Where p is the back pressure, a is the tube radius, and l is its length.

Thus, only where the electroosmotic flow is zero, that is, the so-called stationary level, can the electrophoretic mobility of the particles be measured. To establish the position of the stationary level, let us consider the situation in a microelectrophoresis cell of circular cross as schematically represented in Figure 5.7.

The electroosmotic effects give rise to a solution velocity v_{eo} uniform across the cell cross section, toward the electrode of the same sign as the charge on the cell wall. If the cell is closed, a reverse flow will be set up and there will be no net transport of liquid. This is also the case if the cell is not closed, once the necessary hydrostatic pressure is built-up. The reverse flow follows Poiseuille's law, where the condition of no net liquid transport is set by the following equation,

$$\int_{r=0}^{r=a} 2\pi v r \, dr = 0 \tag{5.26}$$

$$v = v_{eo} - C(a^2 - r^2) \tag{5.27}$$

where C is a constant which from Eqs. (5.25) and (5.26) is given by

$$C = \frac{2v_{eo}}{a^2} \tag{5.28}$$

$$v = v_{eo}\left[\left(\frac{2r^2}{a^2}\right) - 1\right] \tag{5.29}$$

At the cell wall $r = a$ and $v = v_{eo}$ as expected. At the center $r = 0$ and $v = -v_{eo}$, that is, the velocity of the liquid flow is equal in magnitude but opposite in direction to that at the wall. The condition for zero liquid velocity is given by the condition:

$$\frac{2r^2}{a^2} = 1 \tag{5.30}$$

or

$$r = \left(\frac{1}{2}\right)^{1/2} a = 0.707a \tag{5.31}$$

Thus, the stationary level is located at a distance of 0.707 of the radius from the center of the tube or 0.146 of the internal diameter from the wall. By focusing the microscope objective at the top and bottom of the walls of the tube, one can easily locate the position of the stationary levels. The average particle velocity is measured at the top and bottom stationary levels by averaging at least 20 measurements in each direction (the eye piece of the microscope is fitted with a graticule).

For large particles (>1 μm and high density) sedimentation may occur during the measurement. In this case, one can use a rectangular cell and observe the particles horizontally from the side of the glass cell. This is illustrated in Figure 5.8.

The position of the stationary levels within the rectangular cell is more difficult to assign and it depends on the ratio of the two axes a and b in Figure 5.8. When $a/b = \infty$, $v(x = 0) = 0$ when $y/b = 0.5774$ so that the stationary levels are at 0.211 of the cell thickness $2b$ from both the front and back walls. The position of the stationary levels for other values of a/b are given in Table 5.3.

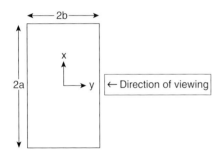

Figure 5.8 Schematic diagram of the rectangular cell.

Table 5.3 Position of stationary levels in rectangular cells.

a/b	(b − y)/2b
∞	0.211
50	0.208
20	0.202
10	0.196

Several commercial instruments for measuring electrophoretic mobility are available (e.g., Rank Brothers, Bottisham Cambridge England, and Pen Kem in USA).

5.3.2
Laser Velocimetry Technique

This method is suitable for small particles that undergo Brownian motion [12]. The scattered light by small particles will show intensity fluctuation as a result of the Brownian diffusion (Doppler shift). When a light beam passes through a colloidal dispersion, an oscillating dipole movement is induced in the particles, thereby radiating the light. Due to the random position of the particles, the intensity of scattered light, at any instant, appear as random diffraction ("speckle" pattern). As the particles undergo Brownian motion, the random configuration of the pattern will fluctuate, such that the time taken for an intensity maximum to become a minimum (the coherence time) corresponds approximately to the time required for a particle to move one wavelength λ. Using a photomultiplier of active area about the diffraction maximum (i.e., one coherent area) this intensity fluctuation can be measured. The analog output is digitized (using a digital correlator) that measures the photocount (or intensity) correlation function of scattered light. The intensity fluctuation is schematically illustrated in Figure 5.9.

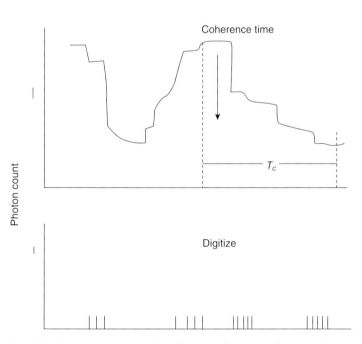

Figure 5.9 Schematic representation of intensity fluctuation of scattered light.

The photocount correlation function $G^{(2)}(\tau)$ is given by

$$g^{(2)} = B[1+\gamma^2 g^{(1)}(\tau)]^2 \tag{5.32}$$

where τ is the correlation delay time.

The correlator compares $g^{(2)}(\tau)$ for many values of τ. B is the Background value to which $g^{(2)}(\tau)$ decays at long delay times. $g^{(1)}(\tau)$ is the normalized correlation function of the scattered electric field and γ is a constant (~1).

For monodispersed noninteracting particles,

$$g^{(1)}(\tau) = \exp(-\Gamma\gamma) \tag{5.33}$$

Γ is the decay rate or inverse coherence time that is related to the translational diffusion coefficient D:

$$\Gamma = DK^2 \tag{5.34}$$

where K is the scattering vector:

$$K = \left(\frac{4\pi n}{\lambda_0}\right)\sin\left(\frac{\theta}{2}\right) \tag{5.35}$$

The particle radius R can be calculated from D using the Stokes–Einstein equation:

$$D = \frac{kT}{6\pi\eta_0 R} \tag{5.36}$$

where η_0 is the viscosity of the medium.

If an electric field is placed at right angles to the incident light and in the plane defined by the incident and observation beam, the line broadening is unaffected but the center frequency of the scattered light is shifted to an extent determined by the electrophoretic mobility. The shift is very small compared to the incident frequency (~100 Hz for and incident frequency of ~6×10^{14} Hz) but with a laser source it can be detected by heterodyning (i.e., mixing) the scattered light with the incident beam and detecting the output of the difference frequency. The homodyne method may be applied in which case a modulator to generate an apparent Doppler shift at the modulated frequency is used. To increase the sensitivity of the laser Doppler method, the electric fields are much higher than those used in conventional electrophoresis. The Joule heating is minimized by pulsing of the electric field in opposite directions. The Brownian motion of the particles also contributes to the Doppler shift and an approximate correction can be made by subtracting the peak width obtained in the absence of an electric field from the electrophoretic spectrum. A He–Ne Laser is used as the light source and the output of the laser is split into two coherent beams that are cross-focused in the cell to illuminate the sample. The light scattered by the particle, together with the reference beam is detected by a photomultiplier. The output is amplified and analyzed to transform the signals to a frequency distribution spectrum. At the intersection of the beam interferences of known spacing are formed.

The magnitude of the Doppler shift Δv is used to calculate the electrophoretic mobility u using the following expression:

$$\Delta v = \left(\frac{2n}{\lambda_0}\right)\sin\left(\frac{\theta}{2}\right)uE \qquad (5.37)$$

where n is the refractive index of the medium, λ_0 is the incident wavelength in vacuum, θ is the scattering angle, and E is the field strength.

Several commercial instruments are available for measuring the electrophoretic light scattering: (i) the Colter DELSA 440SX (Colter Corporation, USA) is a multi-angle laser Doppler system employing heterodyning and autocorrelation signal processing. Measurements are made at four scattering angles (8° 17°, 25°, and 34°) and the temperature of the cell is controlled by a Peltier device. The instrument reports the electrophoretic mobility, zeta potential, conductivity and particle size distribution. (ii) Malvern (Malvern Instruments, UK) has two instruments: The ZetaSizer 3000 and ZetaSizer 5000: The ZetaSizer 3000 is a laser Doppler System using crossed beam optical configuration and homodyne detection with photon correlation signal processing. The zeta potential is measured using laser Doppler velocimetry and the particle size is measured using photon correlation spectroscopy (PCS). The ZetaSizer 5000 uses PCS to measure both (i) movement of the particles in an electric field for zeta potential determination and (ii) random diffusion of particles at different measuring angles for size measurement on the same sample. The Manufacturer claims that zeta potential for particles in the range 50 nm to 30 µm can be measured. In both instruments, a Peltier device is used for temperature control.

5.4
Electroacoustic Methods

The mobility of a particle in an alternating field is termed dynamic mobility, to distinguish it from the electrophoretic mobility in a static electric field described above [13]. The principle of the technique is based on the creation of an electric potential by a sound wave transmitted through an electrolyte solution, as described by Debye [14]. The potential, termed the ionic vibration potential (IVP), arises from the difference in the frictional forces and the inertia of hydrated ions subjected to ultrasound waves. The effect of the ultrasonic compression is different for ions of different masses and the displacement amplitudes are different for anions and cations. Hence, the sound waves create periodically changing electric charge densities. This original theory of Debye was extended to include electrophoretic, relaxation, and pressure gradient forces [15, 16].

A much stronger effect can be observed in colloidal dispersions. The sound waves transmitted by the suspension of charged particles generate an electric field because the relative motion of the two phases is different. The displacement of a charged particle from its environment by the ultrasound waves generates an alternating potential, termed colloidal vibration potential (CVP). The IVP and CVP are both called ultrasound vibration potential (UVP).

The converse effect, namely the generation of sound waves by an alternating electric field [17] in a colloidal dispersion can be measured and is termed the electrokinetic sonic amplitude (ESA). The theory for the ESA effect has been developed by O'Brian and coworkers [18–22]. The dynamic mobility can be determined by measuring either UVP or ESA, although in general the ESA is the preferred method. Several commercial instruments are available for measurement of the dynamic mobility: (i) the ESA-8000 system from Matec Applied Sciences that can measure both CVP and ESA signals; (ii) the Pen Kem System 7000 Acoustophoretic titrator that measures the CVP, conductivity, pH, temperature, pressure amplitude and sound velocity.

In the ESA system (from Matec) and the Acoustosizer (from colloidal dynamics) the dispersion is subjected to a high frequency alternating field and the ESA signal is measured. The ESA-8000 operates at constant frequency of ~1 MHz and the dynamic mobility and zeta potential (but not particle size) are measured. The acoustosizer that operates at various frequencies of the applied electric field can measure the particle mobility, zeta potential, and particle size.

The frequency synthesizer feeds a continuous sinusoidal voltage into a grated amplifier that creates a pulse of sinusoidal voltage across the electrodes in the dispersion. The pulse generates sound waves that appear to emanate from the electrodes. The oscillation, the back-and-forth movement of the particle caused by an electric field is the product of the particle charge times the applied field strength. When the direction of the field is alternating, particles in the suspension between the electrodes are driven away toward the electrodes. The magnitude and phase angle of the ESA signal created is measured with a piezoelectric transducer mounted on a solid nonconductive (glass) rod attached to the electrode as illustrated in Figure 5.10. The purpose of this nonconductive acoustic delay line is to separate the transducer from the high-frequency electric field in the cell. Three pulses of the voltage signal are recorded as schematically shown in Figure 5.11. The first pulse of the signal, shown on the left, is generated when the voltage pulse is applied to the sample and is unrelated to the ESA effect. This first pulse of the

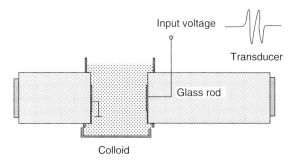

Figure 5.10 Schematic representation of the AcoustoSizer cell.

Figure 5.11 Signals from the right-hand transducer.

signal is received before the sound has sufficient time to pass down the glass rod and is an electronic cross-talk deleted from data processing. The second and third pulses are ESA signals. The second pulse is detected by the nearest electrode. This pulse is used for data processing to determine the particle size and zeta potential. The third pulse originates from the other electrode and is deleted.

In addition to the electrodes, the sample cell of the ESA instruments also houses sensors for pH, conductivity, and temperature measurements. It is also equipped with a stirrer and the system is linked to a digital titrator for dynamic mobility and zeta potential measurements as a function of pH.

To convert the ESA signal to dynamic mobility one needs to know the density of the disperse phase and dispersion medium, the volume fraction of the particles and the velocity of sound in the solvent. As shown earlier, to convert mobility to zeta potential one needs to know the viscosity of the dispersion medium and its relative permittivity. Because of the inertia effects in dynamic mobility measurements, the weight average particle size has to be known.

For dilute suspensions with a volume fraction $\phi = 0.02$, the dynamic mobility u_d can be calculated from the electrokinetic sonic amplitude $A_{ESA}(\omega)$ using the following expression [18, 19]:

$$A_{ESA}(\omega) = Q(\omega)\varphi(\Delta\rho/\rho)(u_d) \tag{5.38}$$

Where ω is the angular frequency of the applied field, $\Delta\rho$ is the density difference between the particle (with density ρ) and the medium. $Q(\omega)$ is an instrument-related coefficient independent of the system being measured.

For a dilute dispersion of spherical particles with $\phi < 0.1$, a thin double layer ($\kappa R > 50$) and narrow particle size distribution (with standard deviation < 20% of the mean size), u_d can be related to the zeta potential ζ by the following equation [18]:

$$u_d = \frac{2\varepsilon\zeta}{3\eta}G\left(\frac{\omega R^2}{v}\right)[1 + f(\lambda, \omega)] \tag{5.39}$$

where ε is the permittivity of the liquid (that is equal to $\varepsilon_r\,\varepsilon_0$, defined before), R is the particle radius, η is the viscosity of the medium, λ is the double layer conductance, and v is the kinematic viscosity ($= \eta/\rho$). G is a factor that represents particle inertia, which reduces the magnitude of u_d and increases the phase lag in a monotonic fashion as the frequency increases. This inertia factor can be used to calculate the particle size from electroacoustic data. The factor $[1 + f(\lambda,\omega)]$ is proportional to the tangential component of the electric field and dependent on the particle permittivity and a surface conductance parameter λ. For most suspensions with large κR, the effect of surface conductance is insignificant and the particle permittivity/liquid permittivity $\varepsilon_p/\varepsilon$ is small. In most cases where the ionic strength is at least $10^{-3}\,\text{mol dm}^{-3}$ and a zeta potential $<75\,\text{mV}$, the factor $[1 + f(\lambda,\omega)]$ assumes the value 0.5. In this case, the dynamic mobility is given by the simple expression,

$$u_d = \frac{\varepsilon\zeta}{\eta}G(\alpha) \tag{5.40}$$

Equation (5.40) is identical to the Smoluchowski equation (5.17), except for the inertia factor $G(\alpha)$.

The equation for converting the ESA amplitude, A_{ESA}, to dynamic mobility is given by

$$u_d = \frac{A_{ESA}}{\varphi v_s \Delta\rho}G(\alpha)^{-1} \tag{5.41}$$

The zeta potential ζ is given by

$$\zeta = \frac{u_d \eta}{\varepsilon}G(\alpha)^{-1} = \frac{A_{ESA}}{\varphi v_s \Delta\rho}G(\alpha)^{-1} \tag{5.42}$$

For a polydisperse system $\langle u_d \rangle$ is given by

$$\langle u_d(\omega)\rangle = \int_0^\infty u(\omega, R)p(R)\,dR \tag{5.43}$$

where $u(\omega,R)$ is the average dynamic mobility of particles with radius R at a frequency ω, and $pR\,dR$ is the mass fraction of particles with radii in the range $R \pm dR/2$.

The ESA measurements can also be applied for determining the particle size in a suspension from particle mobilities. The electric force acting upon a particle is opposed by the hydrodynamic friction and inertia of the particles. At low frequencies of alternating electric field, the inertial force is insignificant and the particle moves in the alternating electric field with the same velocity as it would have moved in a constant field. The particle mobility at low frequencies can be measured to calculate the zeta potential. At high frequencies, the inertia of the particle increases causing the velocity of the particle to decrease and the movement of the particle to lag behind the field. This is illustrated in Figure 5.12 that shows the variation of applied field and particle velocity with time. Since inertia depends on particle mass, both of these effects depend on the particle mass and consequently on its size. Hence, the both zeta potential and particle size can be determined from the ESA signal, if the frequency of the alternating field is sufficiently high. This is the method that is provided by the acoustosizer from colloidal dynamics.

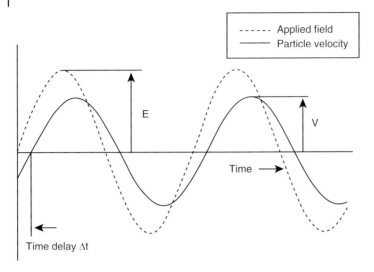

Figure 5.12 Variation of applied field and particle velocity with time at high frequency.

Several variables affect the ESA measurements and these are listed as follows.

i) **Particle concentration range:** Very dilute suspensions generate a weak signal and are not suitable for ESA measurements. The magnitude of the ESA signal is proportional to the average particle mobility, the volume fraction of the particles ϕ, and the density difference between the particles and the medium $\Delta\rho$. To obtain a signal that is at least one order of magnitude higher than the background electrical noise (~0.002 mPa M V^{-1}) the concentration and/or the density difference have to be sufficiently large. If the density difference between the particles and medium is small, for example, polystyrene latex with $\Delta\rho \sim 0.05$ then a sufficiently high concentration ($\phi > 0.02$) is needed to obtain a reasonably strong ESA signal. The accuracy of the ESA measurement is also not good at high ϕ values. This is due to the nonlinearity of the ESA amplitude–ϕ relationship at high ϕ values. Such deviation becomes appreciable at $\phi > 0.1$. However, reasonable values of zeta potential can be obtained from ESA measurements up to $\phi = 0.2$. Above this concentration, the measurements are not sufficiently accurate and the results obtained can only be used for qualitative assessment.

ii) **Electrolyte effects:** Ions in the dispersion generate electroacoustic (IVP) potential and the ESP signal is therefore a composite of the signals created by the particles and ions. However, the ionic contribution is relatively small, unless the particle concentration is low, their zeta potential is low and the ionic concentration is high. The ESA system is therefore not suitable for dynamic mobility and zeta potential measurements in systems with electrolyte concentration higher than 0.3 mol dm^{-3} KCl.

iii) **Temperature:** Since the viscosity of the dispersion decreases by ~2% per °C and its conductivity increases by about the same amount, it is important must be accurately controlled using a Peltier device. Temperature control should also be maintained during sample preparation, for example when the suspension is sonicated. To avoid overheating, the sample should be cooled in an ice bath at regular intervals during sonication.

iv) **Calibration and accuracy:** The electroacoustic probe should be calibrated using a standard reference dispersion such as polystyrene latex or colloidal silica (Ludox). The common sources of error are unsuitable particle concentration (too low or too high), irregular particle shape, polydispersity, electrolyte signals, temperature variations, sedimentation, coagulation, and entrained air bubbles. The latter in particular can cause erroneous ESA signal fluctuations resulting from weakening of the sound by the air bubbles. In many cases, the zeta potential results obtained using the ESA method do not agree with those obtained using other methods such as microelectrophoresis or laser velocimetry. However, the difference seldom exceeds 20% and this makes the ESA method more convenient for measurement of many industrial methods. The main advantages are the speed of measurement and the dispersion does not need to be diluted in which the state of the suspension could be changed.

References

1 Hunter, R.J. (1981) *Zeta Potential in Colloid Science, Principles and Applications*, Academic Press, London.
2 Gouy, G. (1910) *J. Phys.*, **9**, 457; (1917) *Ann. Phys.*, **7**, 129.
3 Chapman, D.L. (1913) *Philos. Mag.*, **25**, 475.
4 Stern, O. (1924) *Z. Electrochem.*, **30**, 508.
5 Grahame, D.C. (1947) *Chem. Rev.*, **41**, 44.
6 von Smoluchowski (1914) *Handbuch der Electricitat und des Magnetismus*, vol. II, Barth, Leipzig.
7 Huckel, E. (1924) *Phys. Z.*, **25**, 204.
8 Overbeek, J.Th.G., and Bijesterbosch, B.H. (1979) *Electrokinetic Separation Methods* (eds. P.G. Righetti, C.J. van Oss, and J.W. Vanderhoff), Elsevier/North Holland Biomedical Press, The Netherlands.
9 Henry, D.C. (1948) *Proc. R. Soc., Lond.*, **A133**, 106.
10 Wiersema, P.H., Loeb, A.L., and Overbeek, J.Th.G. (1967) *J. Colloid Interface Sci.*, **22**, 78.
11 Ottewill, R.H., and Shaw, J.N. (1972) *J. Electroanal. Interfacial Electrochem.*, **37**, 133.
12 Pusey, P.N. (1973) *Industrial Polymers: "Characterisation by Molecular Weights"* (eds J.H.S. Green and R. Dietz), Transcripta Books, London.
13 (a) Kissa, E. (1999) *Dispersions: Characterization, Testing and Measurement*, Marcel Dekker, New York; (b) Debye, P. (1933) *J. Chem. Phys.*, **1**, 13.
14 Bugosh, J., Yeager, E., and Hovarka, F. (1947) *J. Chem. Phys.*, **15**, 542.
15 Yeager, E., Bugosh, J., Hovarka, F., and McCarthy, J. (1949) *J. Chem. Phys.*, **17**, 411.
16 Dukhin, A.S., and Goetz, P.J. (1998) *Colloids Surf.*, **144**, 49.
17 Oja, T., Petersen, G.L., and Cannon, D.C. (1985) US Patent 4,497,208.
18 O'Brian, R.W. (1988) *J. Fluid Mech.*, **190**, 71.
19 O'Brian, R.W. (1990) *J. Fluid Mech.*, **212**, 81.
20 O'Brian, R.W., Garaside, P., and Hunter, R.J. (1994) *Langmuir*, **10**, 931.
21 O'Brian, R.W., Cannon, D.W., and Rowlands, W.N. (1995) *J. Colloid Interface Sci.*, **173**, 406.
22 Rowlands, W.N., and O'Brian, R.W. (1995) *J. Colloid Interface Sci.*, **175**, 190.

6
General Classification of Dispersing Agents and Adsorption of Surfactants at the Solid/Liquid Interface

As mentioned in Chapter 1, for dispersing powders into liquids, one usually requires the addition of a dispersing agent that satisfies the following requirements: lowers the surface tension of the liquid to help wetting of the powder; adsorbs at the solid–liquid (S–L) interface to lower the solid–liquid interfacial tension; lowers the contact angle of the liquid on the solid surface (zero contact angle is very common); helps to break-up the aggregates and agglomerates as well as in subdivision of the particles into smaller units; stabilizes the particles formed against any aggregation (or rejoining).

All dispersing agents are surface active and they can be simple surfactants (anionic, cationic, zwitterionic, or nonionic), polymers, or polyelectrolytes. The dispersing agent should be soluble (or at least dispersible) in the liquid medium and it should adsorb at the solid–liquid interface.

In this chapter, we will describe the general classification of dispersing agents. The adsorption of surfactants at the solid–liquid interface is also discussed in this chapter, whereas the adsorption of polymers will be treated in Chapter 7.

6.1
Classification of Dispersing Agents

6.1.1
Surfactants

Three main classes may be distinguished, namely anionic, cationic, and amphoteric. It should be mentioned also that a fourth class of surfactants, usually referred to as polymeric surfactants, has been used for many years for preparation of suspensions and their stabilization.

6.1.2
Anionic Surfactants

A general formula may be ascribed to anionic surfactants as follows:

Carboxylates: $C_nH_{2n+1}COO^-X$
Sulfates: $C_nH_{2n+1}OSO_3^-X$
Sulfonates: $C_nH_{2n+1}SO_3^-X$
Phosphates: $C_nH_{2n+1}OPO(OH)O^-X$

with n being the range 8–16 atoms and the counterion X is usually Na^+.

Several other anionic surfactants are commercially available such as sulphosuccinates, isethionates, and taurates and these are sometimes used for special applications. Anionic surfactants can be modified by incorporating polyethylene oxide chains, for example, dodecyl polyethylene oxide sulfate (ether sulfates).

6.1.3
Cationic Surfactants

The most common cationic surfactants are the quaternary ammonium compounds with the general formula $R'R''R'''R''''N^+X^-$, where X^- is usually chloride ion and R represents alkyl groups. A common class of cationics is the alkyl trimethyl ammonium chloride, where R contains 8–18 C atoms, for example, dodecyl trimethyl ammonium chloride, $C_{12}H_{25}(CH_3)_3NCl$. Cationic surfactants can also be modified by incorporating polyethylene oxide chains, for example, dodecyl methyl polyethylene oxide ammonium chloride. Cationic surfactants are generally water soluble when there is only the one long alkyl group. They are generally compatible with most inorganic ions and hard water. Cationics are generally stable to pH changes, both acid and alkaline.

6.1.4
Amphoteric (Zwitterionic) Surfactants

These are surfactants containing both cationic and anionic groups. The most common amphoterics are the *N*-alkyl betaines that are derivatives of trimethyl glycine $(CH_3)_3NCH_2COOH$ (that was described as betaine). An example of betaine surfactant is lauryl amido propyl dimethyl betaine $C_{12}H_{25}CON(CH_3)_2CH_2COOH$. These alkyl betaines are sometimes described as alkyl dimethyl glycinates.

The main characteristic of amphoteric surfactants is their dependence on the pH of the solution in which they are dissolved. In acid pH solutions, the molecule acquires a positive charge and it behaves like a cationic, whereas in alkaline pH solutions, they become negatively charged and behave like an anionic. A specific pH can be defined at which both ionic groups show equal ionization (the isoelectric point of the molecule). This can be described by the following scheme:

$$N^+\ldots COOH \leftrightarrow N^+\ldots COO^- \leftrightarrow NH\ldots COO^-$$
acid pH < 3   isoelectric pH > 6 alkaline

Amphoteric surfactants are sometimes referred to as zwitterionic molecules. They are soluble in water, but the solubility shows a minimum at the isoelectric point. Amphoterics show excellent compatibility with other surfactants, forming mixed micelles. They are chemically stable both in acids and alkalis. The surface activity of amphoterics varies widely and it depends on the distance between the charged groups and they show a maximum in the surface activity at the isoelectric point. Another class of amphoteric is the *N*-alkyl amino propionates having the structure R–NHCH$_2$CH$_2$COOH. The NH group is reactive and can react with another acid molecule (e.g., acrylic) to form an amino dipropoinate R–N(CH$_2$CH$_2$COOH)$_2$.

6.1.5
Nonionic Surfactants

The most common nonionic surfactants are those based on ethylene oxide, referred to as ethoxylated surfactants. Several classes can be distinguished: alcohol ethoxylates, alkyl phenol ethoxylates, fatty acid ethoxylates, monoalkaolamide ethoxylates, sorbitan ester ethoxylates, fatty amine ethoxylates, and ethylene oxide–propylene oxide copolymers (sometimes referred to as polymeric surfactants). Amine oxides and sulphinyl surfactants represent nonionics with a small head group.

6.1.6
Alcohol Ethoxylates

These are generally produced by ethoxylation of a fatty chain alcohol such as dodecanol. Several generic names are given to this class of surfactants such as ethoxylated fatty alcohols, alkyl polyoxyethylene glycol, monoalkyl polyethylene oxide glycol ethers, etc. A typical example is dodecyl hexaoxyethylene glycol monoether with the chemical formula C$_{12}$H$_{25}$(OCH$_2$CH$_2$O)$_6$OH (sometimes abbreviated as C$_{12}$E$_6$). In practice, the starting alcohol will have a distribution of alkyl chain lengths and the resulting ethoxylate will have a distribution of ethylene oxide chain length. Thus, the numbers listed in the literature refer to average numbers.

The critical micelle concentration (cmc) of nonionic surfactants is about two orders of magnitude lower than the corresponding anionics with the same alkyl chain length. The solubility of the alcohol ethoxylates depends both on the alkyl chain length and the number of ethylene oxide units in the molecule. Molecules with an average alkyl chain length of 12 C atoms and containing more than 5 EO units are usually soluble in water at room temperature. However, as the temperature of the solution is gradually raised, the solution becomes cloudy (as a result of dehydration of the poly(ethylene oxide) PEO chain) and the temperature at which this occurs is referred to as the cloud point (CP) of the surfactant. At a given alkyl chain length, the CP increases with increase in the EO chain of the molecule. The CP changes with change of concentration of the surfactant solution and the trade literature usually quotes the CP of a 1% solution. The CP is also affected by the presence of electrolyte in the aqueous solution. Most electrolytes lower the CP of a nonionic surfactant solution. Nonionics tend to have the maximum surface activity near to the CP. The CP of most nonionics increases markedly on the addition of small quantities of anionic surfactants. The surface tension of alcohol ethoxylate solutions decreases with decrease in the EO units of the chain. The viscosity of a nonionic surfactant solution increases gradually with increase in its concentration, but at a critical concentration (that depends on the alkyl and EO chain length) the viscosity shows a rapid increase and ultimately a gel-like structure appears. This results from the formation of the liquid crystalline structure of the hexagonal type. In many cases, the viscosity reaches a maximum after which it shows a decrease due to the formation of other structures (e.g., lamellar phases).

6.1.7
Alkyl Phenol Ethoxylates

These are prepared by reaction of ethylene oxide with the appropriate alkyl phenol. The most common surfactants of this type are those based on nonyl phenol. These surfactants are cheap to produce, but they suffer from the problem of biodegradability and potential toxicity (the byproduct of degradation is nonyl phenol that has considerable toxicity). Despite these problems, nonyl phenol ethoxylates are still used in many suspension formulations, due to their advantageous properties, such as their solubility and strong adsorption at the solid–liquid interface.

A more effective nonionic surfactant with a strong adsorption is obtained by using a tristyrylphenol with PEO:

The tristyrylphenol hydrophobic chain adsorbs strongly on a hydrophobic surface by simultaneous adsorption of the styrene chains and the phenyl group (the hydrophobic chain may lie flat on the surface leaving the PEO chain dangling in solution and providing an effective steric barrier).

6.1.8
Fatty Acid Ethoxylates

These are produced by reaction of ethylene oxide with a fatty acid or a polyglycol and they have the general formula $RCOO-(CH_2CH_2O)_nH$. When a polyglycol is used, a mixture of mono- and di-ester $(RCOO-(CH_2CH_2O)_n-OCOR)$ is produced. These surfactants are generally soluble in water provided there is enough EO units and the alkyl chain length of the acid is not too long. The monoesters are much more soluble in water than the di-esters. In the latter case, a longer EO chain is required to render the molecule soluble. The surfactants are compatible with aqueous ions, provided there is not much unreacted acid. However, these surfactants undergo hydrolysis in highly alkaline solutions.

6.1.9
Sorbitan Esters and Their Ethoxylated Derivatives (Spans and Tweens)

The fatty acid esters of sorbitan (generally referred to as Spans, an Atlas commercial trade name) and their ethoxylated derivatives (generally referred to as Tweens) are perhaps one of the most commonly used nonionics. The sorbitan esters are produced by reaction of sorbitol with a fatty acid at a high temperature (>200 °C). The sorbitol dehydrates to 1,4-sorbitan and then esterification takes place. If one mole of fatty acid is reacted with one mole of sorbitol, one obtains a monoester (some di-ester is also produced as a by product). Thus, sorbitan monoester has the following general formula:

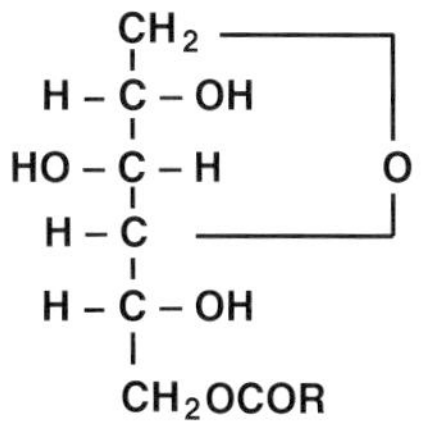

```
        CH2 ─────┐
    H ─ C ─ OH   │
   HO ─ C ─ H    O
    H ─ C ───────┘
    H ─ C ─ OH
        │
       CH2OCOR
```

The free OH groups in the molecule can be esterified, producing di- and tri-esters. Several products are available depending on the nature of the alkyl group of the acid and whether the product is a mono-, di-, or tri-ester. Some examples are given below,

Sorbitan monolaurate – Span 20
Sorbitan monopalmitate – Span 40
Sorbitan monostearate – Span 60
Sorbitan mono-oleate – Span 80
Sorbitan tristearate – Span 65
Sorbitan trioleate – Span 85

The ethoxylated derivatives of Spans (Tweens) are produced by reaction of ethylene oxide on any hydroxyl group remaining on the sorbitan ester group. Alternatively, the sorbitol is first ethoxylated and then esterified. However, the final product has different surfactant properties to the Tweens. Some examples of Tween surfactants are given below:

Polyoxyethylene (20) sorbitan monolaurate – Tween 20
Polyoxyethylene (20) sorbitan monopalmitate – Tween 40
Polyoxyethylene (20) sorbitan monostearate – Tween 60
Polyoxyethylene (20) sorbitan mono-oleate – Tween 80
Polyoxyethylene (20) sorbitan tristearate – Tween 65
Polyoxyethylene (20) sorbitan tri-oleate – Tween 85

The sorbitan esters are insoluble in water, but soluble in most organic solvents (low HLB number surfactants). The ethoxylated products are generally soluble in water and they have relatively high HLB numbers. One of the main advantages of the sorbitan esters and their ethoxylated derivatives is their approval as food additives. They are also widely used in cosmetics and some pharmaceutical suspensions.

6.1.10
Ethoxylated Fats and Oils

A number of natural fats and oils have been ethoxylated, for example, linolin (wool fat) and caster oil ethoxylates. These products are useful for application in suspension formulations, for example, as solubilizers.

6.1.11
Amine Ethoxylates

These are prepared by addition of ethylene oxide to primary or secondary fatty amines. With primary amines both hydrogen atoms on the amine group react with ethylene oxide and therefore the resulting surfactant has the following structure:

$$R-N \begin{matrix} (CH_2CH_2O)_x H \\ \\ (CH_2CH_2O)_y H \end{matrix}$$

The above surfactants acquire a cationic character if the EO units are small in number and if the pH is low. However, at high EO levels and neutral pH they behave very similarly to nonionics. At low EO content, the surfactants are not soluble in water, but become soluble in an acid solution. At high pH, the amine ethoxylates are water soluble provided the alkyl chain length of the compound is not long (usually a C_{12} chain is adequate for reasonable solubility at sufficient EO content.

6.1.12
Polymeric Surfactants

Perhaps the simplest type of a polymeric surfactant is a homopolymer, that is formed from the same repeating units, such as poly(ethylene oxide) or poly(vinyl pyrrolidone). These homopolymers have a little surface activity at the o–w interface, since the homopolymer segments (ethylene oxide or vinylpyrrolidone) are highly water soluble and have little affinity to the interface. However, such homopolymers may adsorb significantly at the S–L interface. Even if the adsorption energy per monomer segment to the surface is small (fraction of kT, where k is the Boltzmann constant and T is the absolute temperature), the total adsorption energy per molecule may be sufficient to overcome the unfavorable entropy loss of the molecule at the S–L interface.

Clearly, homopolymers are not the most suitable dispersants. A small variant is to use polymers that contain specific groups that have high affinity to the surface.

This is exemplified by partially hydrolyzed poly(vinyl acetate) (PVAc), technically referred to as poly(vinyl alcohol) (PVA). The polymer is prepared by partial hydrolysis of PVAc, leaving some residual vinyl acetate groups. Most commercially available PVA molecules contain 4–12% acetate groups. These acetate groups that are hydrophobic give the molecule its amphipathic character. On a hydrophobic surface such as polystyrene, the polymer adsorbs with preferential attachment of the acetate groups on the surface, leaving the more hydrophilic vinyl alcohol segments dangling in the aqueous medium.

The most convenient polymeric surfactants are those of the block and graft copolymer type. A block copolymer is a linear arrangement of blocks of variable monomer composition. The nomenclature for a diblock is poly-A-block-poly-B and for a triblock is poly-A-block-poly-B-poly-A. One of the most widely used triblock polymeric surfactants are the "Pluronics" (BASF, Germany), which consists of two poly-A blocks of poly(ethylene oxide) (PEO) and one block of poly(propylene oxide) (PPO).

$$\text{HO}(CH_2CH_2O)_n\text{–}(CH_2CHO)_m\text{–}(CH_2CH_2)_n\text{OH} \text{ abbreviated } (EO)_n(PO)_m(EO)_n$$
$$|$$
$$CH_3$$

Various molecules are available, where n and m are varied systematically.

Trifunctional products are also available where the starting material is glycerol. These have the structure:

$$CH_2\text{–}(PO)_m(EO)_n$$
$$|$$
$$CH\text{–}(PO)_n(EO)_n$$
$$|$$
$$CH_2\text{–}(PO)_m(EO)_n$$

Tetrafunctional products are available where the starting material is ethylene diamine. These have the structures,

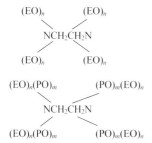

The above polymeric triblocks can be applied as dispersants, whereby the assumption is made that the hydrophobic PPO chain resides at the hydrophobic surface, leaving the two PEO chains dangling in the aqueous solution and hence providing steric repulsion. Although these triblock polymeric surfactants have been widely used in various applications suspensions, some doubt has arisen on how effective these can be. It is generally accepted that the PPO chain is not sufficiently hydrophobic to provide a strong "anchor" to a hydrophobic surface.

Several other di- and triblock copolymers have been synthesized, although these are of limited commercial availability. Typical examples are diblocks of polystyrene-block-polyvinyl alcohol, triblocks of poly(methyl methacrylate)-block poly(ethylene oxide)-block poly(methyl methacrylate), diblocks of polystyrene block-polyethylene oxide and triblocks of polyethylene oxide-block polystyrene-polyethylene oxide. An alternative (and perhaps more efficient) polymeric surfactant is the amphipathic graft copolymer consisting of a polymeric backbone B (polystyrene or polymethyl methacrylate) and several A chains ("teeth") such as polyethylene oxide. This graft copolymer is sometimes referred to as a "comb" stabilizer. This copolymer is usually prepared by grafting a macromonomer such as methoxy polyethylene oxide methacrylate with polymethyl methacrylate. The "grafting onto" technique has also been used to synthesize polystyrene–polyethylene oxide graft copolymers.

Recently, graft copolymers based on polysaccharides have been developed for stabilization of disperse systems. One of the most useful graft copolymers are those based on inulin that is obtained from chicory roots. It is a linear polyfructose chain with a glucose end. When extracted from chicory roots, inulin has a wide range of chain lengths ranging from 2 to 65 fructose units. It is fractionated to obtain a molecule with narrow molecular weight distribution with a degree of polymerization >23 and this is commercially available as INUTEC®N25. The latter molecule is used to prepare a series of graft copolymers by random grafting of alkyl chains (using alky isocyanate) on the inulin backbone. The first molecule of this series is INUTEC®SP1 (Beneo-Remy, Belgium) that is obtained by random grafting of C_{12} alkyl chains. It has an average molecular weight of ~ 5000 Da and its structure is given in Figure 6.1. The molecule is schematically illustrated in Figure 6.2 that shows the hydrophilic polyfructose chain (backbone) and the randomly attached alky chains. The main advantages of INUTEC®SP1 as stabilizer for disperse systems are: (i) Strong adsorption to the particle or droplet by multipoint attachment with several alky chains. This ensures lack of desorption

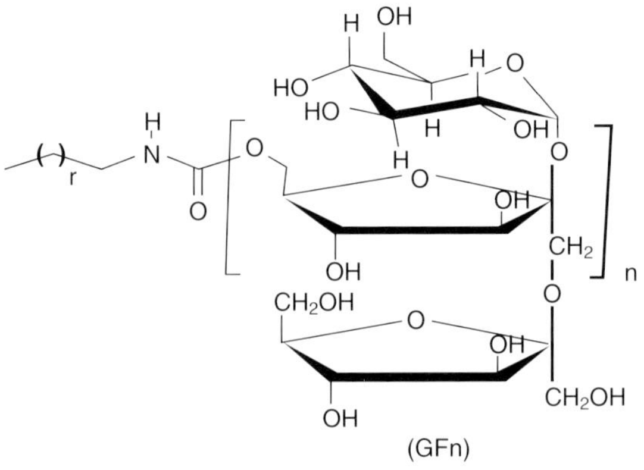

Figure 6.1 Structure of INUTEC®SP1.

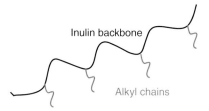

Figure 6.2 Schematic representation of INUTEC®SP1 polymeric surfactant.

and displacement of the molecule from the interface. (ii) Strong hydration of the linear polyfructose chains both in water and in the presence of high electrolyte concentrations and high temperatures. This ensures effective steric stabilization (see below).

6.1.13
Polyelectrolytes

A good example is sulfonated alkyl naphthalene formaldehyde condensates,

$$\text{(NaSO}_3\text{)} \quad R - \text{naphthyl} - CH_2 - \text{naphthyl} - R \quad (SO_3Na)n$$

The molecule has a wide distribution of molecular weights. The polyelectrolyte is a good dispersant for many hydrophobic solids and in some cases a wetter is not needed since the low molecular weight species can diffuse fast to the interface.

6.1.14
Adsorption of Surfactants at the Solid–Liquid Interface

The adsorption of ionic surfactants at the solid–liquid interface is of major technological importance. Apart from controlling the colloid stability as previously discussed in Chapter 4, ionic surfactants are also used as wetting agents for hydrophobic solids in aqueous media. This was discussed in detail in Chapters 2 and 3.

Surfactant adsorption is relatively simpler than polymer adsorption. This stems from the fact that surfactants consist of a small number of units and they mostly are reversibly adsorbed, allowing one to apply thermodynamic treatments. In this case, it is possible to describe the adsorption in terms of the various interaction parameters, namely chain–surface, chain–solvent, and surface–solvent. Moreover, the conformation of the surfactant molecules at the interface can be deduced from these simple interactions parameters. However, in some cases the interaction parameters may involve ill-defined forces, such as hydrophobic bonding, solvation forces, and chemisorption. In addition, the adsorption of ionic surfactants involves electrostatic forces particularly with polar surfaces containing ionogenic groups. For that reason, the adsorption of ionic and nonionic surfactants will be treated separately. The surfaces (substrates) can be also hydrophobic or hydrophilic and these may be treated separately.

6.1.15
Adsorption of Ionic Surfactants on Hydrophobic Surfaces

The adsorption of ionic surfactants on hydrophobic surfaces such as carbon black, polymer surfaces, and ceramics (silicon carbide or silicon nitride) is governed by hydrophobic interaction between the alkyl chain of the surfactant and the hydrophobic surface. In this case, electrostatic interaction will play a relatively smaller role. However, if the surfactant head group is of the same sign of charge as that on the substrate surface, electrostatic repulsion may oppose adsorption. In contrast, if the head groups are of opposite sign to the surface, adsorption may be enhanced. Since the adsorption depends on the magnitude of the hydrophobic bonding free energy, the amount of surfactant adsorbed increases directly with increase in the alkyl chain length in accordance with Traube's rule.

The adsorption of ionic surfactants on hydrophobic surfaces may be represented by the Stern–Langmuir isotherm [1]. The main assumptions used in the derivation of the Stern–Langmuir equation are: (i) Only one type of ion, the surfactant ion, is specifically adsorbed. This assumption is reasonable at low surface coverage in the presence of electrolyte, which is considered as "indifferent." (ii) One surfactant ion replaces one solvent molecule from the surface, that is, ion and solvent molecules are of the same size. (iii) The surface is considered homogeneous. (iv) Dipole terms and lateral chain–chain interactions are absent.

Consider a substrate containing N_s sites (mol m^{-2}) on which Γ moles m^{-2} of surfactant ions are adsorbed. The surface coverage θ is (Γ/N_s) and the fraction of uncovered surface is $(1 - \theta)$. The rate of adsorption is proportional to the surfactant concentration expressed in mole fraction, $(C/55.5)$, and the fraction of free surface $(1 - \theta)$, that is,

$$\text{Rate of adsorption} = k_{\text{ads}}\left(\frac{C}{55.5}\right)(1-\theta) \tag{6.1}$$

where k_{ads} is the rate constant for adsorption.

The rate of desorption is proportional to the fraction of surface covered θ:

$$\text{Rate of desorption} = k_{\text{des}}\theta \tag{6.2}$$

At equilibrium, the rate of adsorption is equal to the rate of desorption and the ratio of $(k_{\text{ads}}/k_{\text{des}})$ is the equilibrium constant K, that is,

$$\frac{\theta}{(1-\theta)} = \frac{C}{55.5}K \tag{6.3}$$

The equilibrium constant K is related to the standard free energy of adsorption by

$$-\Delta G_{ads}^0 = RT \ln K \tag{6.4}$$

R is the gas constant and T is the absolute temperature. Equation (6.4) can be written in the form
or

$$K = \exp\left(-\frac{\Delta G_{ads}^0}{RT}\right) \tag{6.5}$$

Combining Eqs. (6.3) and (6.5),

$$\frac{\theta}{1-\theta} = \frac{C}{55.5} \exp\left(-\frac{\Delta G_{ads}^0}{RT}\right) \tag{6.6}$$

Equation (6.6) applies only at low surface coverage ($\theta < 0.1$) where lateral interaction between the surfactant ions can be neglected.

At high surface coverage ($\theta > 0.1$) one should take the lateral interaction between the chains into account, by introducing a constant A, for example, using the Frumkin–Fowler–Guggenheim equation (6.1):

$$\frac{\theta}{(1-\theta)} \exp(A\theta) = \frac{C}{55.5} \exp\left(-\frac{\Delta G_{ads}^0}{RT}\right) \tag{6.7}$$

Various authors [2, 3] have used the Stern–Langmuir equation in a simple form to describe the adsorption of surfactant ions on mineral surfaces:

$$\Gamma = 2rC \exp\left(-\frac{\Delta G_{ads}^0}{RT}\right) \tag{6.8}$$

Various contributions to the adsorption free energy may be envisaged. To a first approximation, these contributions may be considered to be additive. In the first instance, ΔG_{ads} maybe taken to consist of two main contributions, that is,

$$\Delta G_{ads} = \Delta G_{elec} + \Delta G_{spec} \tag{6.9}$$

where ΔG_{elec} accounts for any electrical interactions and ΔG_{spec} is a specific adsorption term that contains all contributions to the adsorption free energy that are dependent on the "specific" (nonelectrical) nature of the system [4]. The electrical contribution ΔG_{elec} is ascribed totally to Coulombic interactions, although it may contain a contribution due to dipole interaction. When the surfactant ions are counterions to the net charge density, for example, a positive surfactant ion on a negative surface, the electrical interaction promotes adsorption. In contrast if the surfactant ions are of the same charge as the surface, for example, a negative surfactant ion on a negative surface, the electrical interaction oppose adsorption.

Several authors subdivided ΔG_{spec} into supposedly separate independent interactions [4, 5], for example,

$$\Delta G_{spec} = \Delta G_{cc} + \Delta G_{cs} + \Delta G_{hs} + \cdots \tag{6.10}$$

where ΔG_{cc} is a term that accounts for the cohesive chain–chain interaction between the hydrophobic moieties of the adsorbed ions, ΔG_{cs} is the term for chain/

substrate interaction whereas ΔG_{hs} is a term for the head group–substrate interaction. ΔG_{cs} can be accounted for by hydrophobic bonding or van der Waals dispersion interactions. The driving force for hydrophobic interactions is principally entropic, arising from the destruction of short-lived structures of water molecules organized around apolar substrates, similar to the process of surfactant micellization. Other physical components of ΔG_{cs} can be envisaged depending on the structure of the adsorbent and adsorbate. For example, the adsorption of a surfactant possessing an alkylaryl hydrophobic residue may involve ion-dipole (or dipole)-induced dipole interaction as a result of π-electron polarization of the aromatic group on close approach to the surface.

Since there is no rigorous theory that can predict adsorption isotherms, the most suitable method to investigate adsorption of surfactants is to determine the adsorption isotherm. Measurement of surfactant adsorption is fairly straightforward. A known mass m (g) of the particles (substrate) with known specific surface area A_s (m^2 g^{-1}) is equilibrated at constant temperature with surfactant solution with initial concentration C_1. The suspension is kept stirred for sufficient time to reach equilibrium. The particles are then removed from the suspension by centrifugation and the equilibrium concentration C_2 is determined using a suitable analytical method. The amount of adsorption Γ (mole m^{-2}) is calculated as follows:

$$\Gamma = \frac{(C_1 - C_2)}{mA_s} \tag{6.11}$$

The adsorption isotherm is represented by plotting Γ versus C_2. A range of surfactant concentrations should be used to cover the whole adsorption process, that is, from the initial low values to the plateau values. To obtain accurate results, the solid should have a high surface area (usually $> 1\,\text{m}^2$).

Several examples may be quoted from the literature to illustrate the adsorption of surfactant ions on solid surfaces. For a model hydrophobic surface, carbon black has been chosen [6, 7]. Figure 6.3 shows typical results for the adsorption of

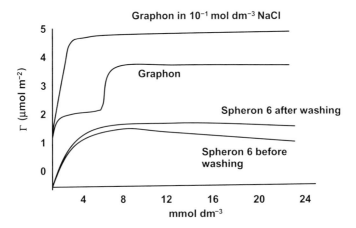

Figure 6.3 Adsorption isotherms for sodium dodecyl sulfate on carbon substrates.

sodium dodecyl sulfate (SDS) on two carbon black surfaces, namely Spheron 6 (untreated) and graphon (graphitized) that also describes the effect of surface treatment. The adsorption of SDS on untreated Spheron 6 tends to show a maximum that is removed on washing. This suggests the removal of impurities from the carbon black which becomes extractable at high surfactant concentration. The plateau adsorption value is $\sim 2 \times 10^{-6}\,mol\,m^{-2}$ ($\sim 2\,\mu mole\,m^{-2}$). This plateau value is reached at $\sim 8\,mmol\,dm^{-3}$ SDS, that is, close to the cmc of the surfactant in the bulk solution. The area per surfactant ion in this case is $\sim 0.7\,nm^2$. Graphitization (graphon) removes the hydrophilic ionizable groups (e.g., –C=O or –COOH), producing a surface that is more hydrophobic. The same occurs by heating Spheron 6 to 2700 °C. This leads to a different adsorption isotherm (Figure 6.3) showing a step (inflection point) at a surfactant concentration in the region of $\sim 6\,mmol\,dm^{-3}$. The first plateau value is $\sim 2.3\,\mu mole\,m^{-2}$ whereas the second plateau value (that occurs at the CMC of the surfactant) is $\sim 4\,\mu mole\,m^{-2}$. It is likely in this case that the surfactant ions adopt different orientations at the first and second plateaus. In the first plateau region, a more "flat" orientation (alkyl chains adsorbing parallel to the surface) is obtained whereas at the second plateau vertical orientation is more favorable, with the polar head groups being directed towards the solution phase. Addition of electrolyte ($10^{-1}\,mol\,dm^{-3}$ NaCl) enhances the surfactant adsorption. This increase is due to the reduction of lateral repulsion between the sulfate head groups and this enhances the adsorption.

The adsorption of ionic surfactants on hydrophobic nonpolar surfaces resembles that for carbon black [8, 9]. For example, Saleeb and Kitchener [8] found similar limiting area for cetyltrimethyl ammonium bromide on graphon and polystyrene ($\sim 0.4\,nm^2$). As with carbon black, the area per molecule depends on the nature and amount of added electrolyte. This can be accounted for in terms of reduction of head group repulsion and/or counterion binging.

Surfactant adsorption close to the CMC may appear Langmuirian, although this does not automatically imply a simple orientation. For example, rearrangement from horizontal to vertical orientation or electrostatic interaction and counterion binding may be masked by simple adsorption isotherms. It is essential, therefore, to combine the adsorption isotherms with other techniques such as microcalorimetry and various spectroscopic methods to obtain a full picture on surfactant adsorption.

6.1.16
Adsorption of Ionic Surfactants on Polar Surfaces

The adsorption of ionic surfactants on polar surfaces that contain ionizable groups may show characteristic features due to additional interaction between the head group and substrate and/or possible chain–chain interaction. This is best illustrated by the results of adsorption of sodium dodecyl sulfonate (SDSe) on alumina at pH = 7.2 obtained by Fuerestenau [10] and shown in Figure 6.4. At the pH value, the alumina is positively charged (the isoelectric point of alumina is at pH $\sim$ 9) and the counterions are Cl^- from the added supporting electrolyte. In Figure 6.4,

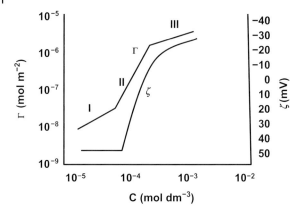

Figure 6.4 Adsorption isotherm for sodium dodecyl sulfonate on alumina and corresponding zeta (ζ).

the saturation adsorption Γ_1 is plotted versus equilibrium surfactant concentration C_1 in logarithmic scales. The figure also shows the results of zeta potential (ζ) measurements (that are measure of the magnitude sign of charge on the surface). Both adsorption and zeta potential results show three distinct regions. The first region that shows a gradual increase of adsorption with increase in concentration, with virtually no change in the value of the zeta potential corresponds to an ion-exchange process [11]. In other words, the surfactant ions simply exchange with the counterions (Cl$^-$) of the supporting electrolyte in the electrical double layer. At a critical surfactant concentration, the adsorption increases dramatically with further increase in surfactant concentration (region II). In this region, the positive zeta potential gradually decrease, reaching a zero value (charge neutralization) after which a negative value is obtained that increases rapidly with increase in surfactant concentration. The rapid increase in region II was explained in terms of "hemimicelle formation" that was originally postulated by Gaudin and Fuertsenau [12]. In other words, at a critical surfactant concentration (to be denoted the CMC of "hemimicelle formation" or better the critical aggregation concentration (CAC)), the hydrophobic moieties of the adsorbed surfactant chains are "squeezed out" from the aqueous solution by forming two-dimensional aggregates on the adsorbent surface. This is analogous to the process of micellization in the bulk solution. However, the CAC is lower than the CMC, indicating that the substrate promotes surfactant aggregation. At a certain surfactant concentration in the hemimicellization process, the isoelectric point is exceeded and, thereafter, the adsorption is hindered by the electrostatic repulsion between the hemimicelles and hence the slope of the adsorption isotherm is reduced (region III).

6.1.17
Adsorption of Nonionic Surfactants

Several types of nonionic surfactants exist, depending on the nature of the polar (hydrophilic) group. The most common type is that based on a poly(oxyethylene) glycol group, that is, $(CH_2CH_2O)_nOH$ (where n can vary from as little as two units to as high as 100 or more units) linked either to an alkyl (C_xH_{2x+1}) or alkyl phenyl $(C_xH_{2x+1}–C_6H_4–)$ group. These surfactants may be abbreviated as C_xE_n or $C_x\phi E_n$ (where C refers to the number of C atoms in the alkyl chain, ϕ denotes C_6H_4 and E denotes ethylene oxide). These ethoxylated surfactants are characterized by a relatively large head group compared to the alkyl chain (when $n > 4$). However, there are nonionic surfactants with small head group such as amine oxides ($-N \rightarrow 0$) head group, phosphate oxide ($-P \rightarrow 0$) or sulphinyl-alkanol ($-SO-(CH_2)_n-OH$) [13]. Most adsorption isotherms in the literature are based on the ethoxylated type surfactants.

In comparing various nonionic surfactants, and in particular those bases on poly(ethylene oxide) (PEO), it is useful to use the "hydrophilic-lipophilic-balance" (HLB) concept that is simply given the relative proportion of the hydrophilic (PEO and OH) and lipophilic (alkyl chain) components. The HLB is simply given by the weight percent of PEO and OH divided by 5. For example for a nonionic surfactant of $C_{12}H_{25}-O-(CH_2-CH_2-O)_4-H$, the HLB is 10.5.

The adsorbents used for studying nonionic surfactant adsorption range from apolar surfaces such as C black, organic pigments, or polystyrene (low energy solids) to polar surfaces such as oxides or silicates (high-energy solids).

The adsorption isotherm of nonionic surfactants are in many cases Langmuirian, like those of most other highly surface active solutes adsorbing from dilute solutions and adsorption is generally reversible. However, several other adsorption types are produced [13] and those are illustrated in Figure 6.5. The steps in the isotherm may be explained in terms of the various adsorbate–adsorbate, adsorbate–adsorbent, and adsorbate–solvent interactions. These orientations are schematically illustrated in Figure 6.6.

In the first stage of adsorption (denoted by I in Figures 6.5 and 6.6), surfactant–surfactant interaction is negligible (low coverage) and adsorption occurs mainly by van der Waals interaction. On a hydrophobic surface, the interaction is dominated by the hydrophobic portion of the surfactant molecule. This is mostly the

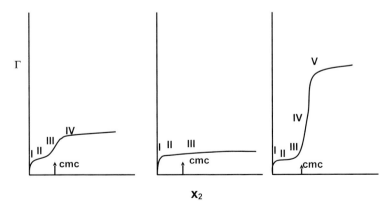

Figure 6.5 Adsorption isotherms corresponding to the three adsorption sequences shown in Figure 6.6.

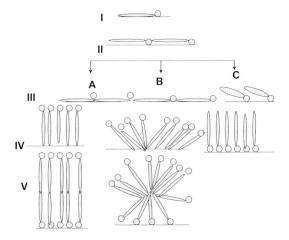

Figure 6.6 Model for adsorption of nonionic surfactants.

case with pharmaceuticals and agrochemicals which have hydrophobic surfaces. Nevertheless, the polar groups of the surfactant may have some interaction with the surface, and the hydrophilic EO groups can have a slight positive adsorption even on a nonpolar adsorbent. When the interaction is due to dispersion forces the heat of adsorption is relatively small and corresponds to the heat liberated by replacing solvent molecules with surfactant. At this stage, the molecule tends to lie flat on the surface because its hydrophobic portion is positively adsorbed, as are also most types of hydrophilic head groups, especially large PEO chains. With the molecule lying parallel to the surface, the adsorption energy will increase in almost equal increments for each additional carbon atom in its alkyl chain, and the initial slope of the isotherm will increase accordingly according to Traube's rule. The same also happens with each additional EO group.

The approach to monolayer saturation with the molecules lying flat (Figure 6.6 II) is accompanied by gradual decrease in the slope of the adsorption isotherm as shown in Figure 6.5. Although most of the "free" solvent molecules will have been displaced from the surface by the time the monolayer is complete, the surfactant molecules themselves will probably stay hydrated at this stage. Increase in the size of the surfactant molecule, for example, by increasing the length of the alkyl chain or the PEO chain will decrease the adsorption (expressed in mol m^{-2}). Increasing temperature should increase the adsorption because dehydration decreases the size of the adsorbate molecules. Increasing temperature reduces the solubility of the nonionic surfactant and this enhances its adsorption (higher surface activity).

The subsequent stages of adsorption are increasingly dominated by adsorbate–adsorbate interaction, although it is the adsorbate–adsorbent interaction that initially determines how the adsorption progresses when stage II is completed. The adsorbate–adsorbent interaction depends on the nature of the adsorbent and the HLB of the surfactant. When the hydrophilic group (e.g., PEO) is only weakly adsorbed it will be displaced from the surface by the alkyl chain of the surfactant molecule (Figure 6.6 III A). This is particularly the case with nonpolar adsorbents such as C black or polystyrene when the surfactant has a short PEO chain (relatively low HLB number). However, if the interaction between the hydrophilic chain (PEO) and polar adsorbent such ass silica or silicates is strong the alkyl chain is displaced as is illustrated in Figure 6.6 III C. The intermediate situation (Figure 6.1 III B) occurs when neither type of displacement is favored and the adsorbate molecules remain flat on the surface.

The change in the amount of adsorption in the third stage (Stage III of Figure 6.5) is unlikely to be large, but as the concentration of the surfactant in the bulk solution approached the CMC there will be a tendency for the alkyl chains of the adsorbed molecules to aggregate. This will cause the molecules to be vertically oriented and there will be a large increase in adsorption (stage IV). The lateral forces due to alkyl chain interactions will compress the head group, and for a PEO chain this will result in a less coiled, more extended conformation. The longer the alkyl chain the greater will be the cohesion force and hence the smaller the surfactant cross-sectional area. This may explain the increase in saturation adsorption with increase in the alkyl chain length and decrease in the number of EO units in the PEO chain. With nonpolar adsorbents the adsorption energy per methylene group is almost the same as the micellization energy, so surface aggregation can occur quite easily even at concentrations below the cmc. However, with polar adsorbents, the head group may be strongly bound to the surface, and partial displacement of a large PEO chain from the surface, needed for close packing, may not be achieved until the surfactant concentration is above the CMC. When the adsorption layer is like that shown in Figure 6.6 IVC the surface becomes hydrophobic.

The interactions occurring in the adsorption layer during the fourth and subsequent stages of adsorption are similar to interactions in bulk solution where enthalpy changes caused by increase alkyl–alkyl interactions balance those due to head group interactions and dehydration process. For this reason, the heat of adsorption remains constant, although adsorption increases with increasing temperature due to dehydration of the head group and its more compact nature.

The parallel between bulk micellization and the surface aggregation process has been emphasized by Klimenko et al. [14, 15] who suggested that above the CMC the adsorbed surfactant molecules form micellar aggregates on the surface as illustrated in Figure 6.6 V. Both hemimicelles and full micelles can be formed on the surface. This picture was supported by Klimenko et al. [14] who found close agreement between saturation adsorption and adsorption calculated based on the assumption that the surface is covered with close-packed hemimicelles.

6.1.18
Theoretical Treatment of Surfactant Adsorption

Kleminko [15, 16] developed a theoretical model for the three stages of adsorption of nonionic surfactants. In the first stage (flat orientation) a modified Langmuir adsorption equation was used:

$$c_2 K_a = \left[\frac{\Gamma_2}{\Gamma_2^* - \Gamma_2(1 - a_1/a_2')a_2'} \frac{a_1}{a_2'} \right] f_2' \qquad (6.12)$$

where c_2 is the equilibrium concentration of surfactant in bulk solution, Γ_2 is the surface excess concentration at c_2, Γ_2^* is the surface excess at the CMC, K_a is a constant, a_1 and a_2' are the effective cross-sectional areas of the solvent and adsorbate molecules in the surface, and f_a' is an adsorbate surface activity coefficient. The term in the square bracket is a type of surface "concentration" that is defined as the ratio of numbers of adsorbed surfactant molecules to the number of solvent molecules in the equilibrium interfacial layer. The constant K_a allows for adsorbate–adsorbent interactions and is therefore related to the energy of adsorption at infinite dilution. The adsorbate surface activity coefficient f_a' accounts for the adsorbate–adsorbate interaction and it is assumed to have the following dependence on Γ_2 in the first stage of adsorption,

$$f_2' = \exp\left[\frac{\Gamma_2}{\Gamma_2^* - \Gamma_2} - \frac{K_2 \Gamma_2}{\Gamma_2^*} \right] \qquad (6.13)$$

where K_2 is an adsorbate–adsorbate interaction constant.

When all the solvent molecules have been displaced and the surface is covered with a close-packed monolayer of horizontal adsorbate molecules, the second stage begins and the EO chains are progressively displaced by alkyl chains of adsorbate molecules. This allows the surface concentration to increase by the following amount:

$$\frac{\Gamma_2 - \Gamma_2'}{\Gamma_2'(a_E'/a_1) - (\Gamma_2 - \Gamma_2')(a_2/a_1)} \qquad (6.14)$$

where Γ_2' is the surface excess at the beginning of the second stage, a_E' is the cross-sectional area of the EO chain lying flat on the surface, and a_2 is the area of the surface covered by each surfactant molecule. The adsorption isotherm for the second stage is given by [14–16]

$$c_2 K_a = \left[\frac{\Gamma_2'}{\Gamma_2^* - \Gamma_2'(1 - a_1/a_2')a_2'} \frac{a_1}{a_2'} + \frac{\Gamma_2 - \Gamma_2'}{\Gamma_2' a_E' - (\Gamma_2 - \Gamma_2')a_2} a_1 \right] f_a'' \qquad (6.15)$$

The activity coefficient f_a'' is assumed to include contributions from EO chain interactions and therefore to differ from f_a' in its dependence on Γ_2. The logarithm of f_a'' is arbitrarily assumed to have linear dependence on Γ_2:

$$\ln f_a'' = \ln(f_a')_{\Gamma_2'} + [\ln(f_a'')_{\Gamma_2^*} - \ln(f_a')_{\Gamma_2'}] \Gamma_2 / (\Gamma_2^* - \Gamma_2') \qquad (6.16)$$

In Eq. (6.16) f_a' with subscript Γ_2', the maximum value of f_a' that is reached when $\Gamma_2 = \Gamma_2'$, is obtained by substituting the value of Γ_2' into Eq. (6.13); f_a'' with subscript Γ_2^*, the maximum value of f_a'' that is reached when $c_2 = c^*$, the CMC, and $\Gamma_2 = \Gamma_2^*$ can be obtained by substitution into Eq. (6.15).

In the final adsorption stage, which starts at the CMC, Kleminko [14–16] assumes that the adsorbed surfactant associates into hemimicelles on the surface. By considering the equilibrium between molecules in the bulk solution and "free" positions in these surface micelles, he drives a simple Langmuir isotherm:

$$c_2 K_a^* = \frac{\Gamma_2}{(\Gamma_2^\infty - \Gamma_2)} \qquad (6.17)$$

where Γ_2^∞ is the maximum surface excess, that is, the surface excess when the surface is covered with close-packed hemimicelles, K_a^* is a constant that is inversely proportional to the cmc because it is assumed that the surface micelles are similar to the bulk solution micelles and c_2 is the equilibrium concentration. Equation (6.17) does not contain an activity coefficient because it is assumed that above the cmc deviations from ideality in the surface and in bulk solution derive from a similar association effect with the result that the two activity coefficients will cancel each other in the adsorption equation.

6.1.19
Examples of Typical Adsorption Isotherms of Model Nonionic Surfactants on Hydrophobic Solids

Corkill *et al.* [17] studied the adsorption of very pure alkyl polyoxyethylene glycol monoethers C_xE_n on graphon (with a nitrogen BET specific surface area of 91 m² g⁻¹). The adsorption isotherms were mostly simple Langmuirian with saturation adsorption reached near or above the cmc. The maximum adsorption increases with increasing the alkyl chain length and decreasing the EO chain length. From the saturation adsorption, the area per molecule at 25 °C was calculated and compared with the value obtained at the solution–air interface (which was obtained from the γ–log c curves and application of the Gibbs adsorption isotherm). The results are given in Table 6.1.

The similarity between the areas at the two interfaces suggests that, at saturation, molecules adsorbed on the solid are vertically oriented as in Figure 6.6 IV. Hexaoxyethylene glycol (E_6) is also positively adsorbed on graphon, indicating some affinity to the carbon surface.

Table 6.1 Area per molecule at the graphon–solution interface and air–solution interface at 25 °C.

Surfactant	Area/molecule (nm²) Graphon–solution interface	Area/molecule (nm²) Air–solution interface
E_6	1.68	–
C_6E_6	0.93	0.94
C_8E_6	0.81	0.83
$C_{10}E_6$	0.65	0.72
$C_{12}E_6$	0.55	0.61
C_8E_9	1.09	1.02
$C_{16}E_9$	–	0.47

The heat of wetting of graphon with solutions of C_8E_6 was measured as a function of surface excess and the results showed an initial linear change in heat with surface coverage corresponding to the replacement of water at the interface by the hydrated surfactant molecules until the surface is saturated with horizontal adsorbate molecules. The plateau value for the heat of immersion is reached at a molecular area of $1.32\,nm^2$ that is close to the cross-sectional area of C_8E_6 lying flat. The constancy of the heat of immersion at higher concentrations is attributed to the balance between decreasing enthalpy, due to elimination of the alkyl–solution interface as the molecules become vertically oriented, and increasing enthalpy associated with the dehydration of the adsorbing molecule. The adsorption thus occurs with a net increase in entropy analogous to the process of micellization.

The adsorption of nonionic surfactants based on poly(ethylene oxide) increases with increase of temperature as is illustrated in Figure 6.7 for C_8E_6 and C_8E_3. Increasing temperature gradually dehydrates the PEO head group and this makes the molecule less hydrophilic and more compact, thus increasing the surface activity and saturation adsorption values. The importance of the surfactant–solvent interaction is apparent on the effect of temperature on the adsorption of C_8E_3. At 4.5 °C the adsorption isotherm is a simple Langmuirian, but at 25 °C and 40 °C there is a very steep rise in adsorption at high surfactant concentrations. This is characteristic of a surface condensation effect and the steep rise occurs at concentrations below those at which surfactant phase separation occurs in bulk solution. These concentrations are shown by broken lines in Figure 6.7.

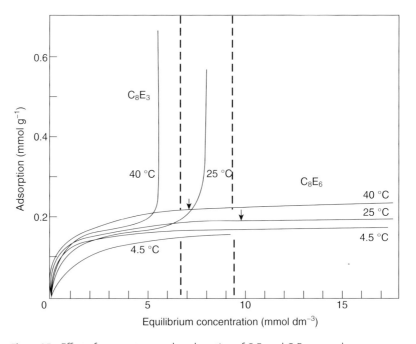

Figure 6.7 Effect of temperature on the adsorption of C_8E_6 and C_8E_3 on graphon.

References

1. Hough, D.B., and Randall, H.M. (1983) *Adsorption from Solution at the Solid/Liquid Interface* (eds G.D. Parfitt and C.H. Rochester), Academic Press, London, p. 247.
2. Fuerstenau, D.W., and Healy, T.W. (1972) *Adsorptive Bubble Seperation Techniques* (ed. R. Lemlich), Academic Press, London, p. 91.
3. Somasundaran, P., and Goddard, E.D. (1979) *Mod. Aspects Electrochem.*, **13**, 207.
4. Healy, T.W. (1974) *J. Macromol. Sci. Chem.*, **118**, 603.
5. Somasundaran, P., and Hannah, H.S. (1979) *Improved Oil Recovery by Surfactant and Polymer Flooding* (eds D.O. Shah and R.S. Schechter), Academic Press, London, p. 205.
6. Greenwood, F.G., Parfitt, G.D., Picton, N.H., and Wharton, D.G. (1968) *Adv. Chem. Ser.*, No. **79**, 135.
7. Day, R.E., Greenwood, F.G., and Parfitt, G.D. (1967) 4th Int. Congress of Surface Active Substances, vol. 18, p. 1005.
8. Saleeb, F.Z., and Kitchener, J.A. (1965) *J. Chem. Soc.*, 911.
9. Conner, P., and Ottewill, R.H. (1971) *J. Colloid Interface Sci.*, **37**, 642.
10. Fuerestenau, D.W. (1971) *The Chemistry of Biosurfaces* (ed. M.L. Hair), Marcel Dekker, New York, p. 91.
11. Wakamatsu, T., and Fuerstenau, D.W. (1968) *Adv. Chem. Ser.*, **71**, 161.
12. Gaudin, A.M., and Fuertsenau, D.W. (1955) *Trans. AIME*, **202**, 958.
13. Clunie, J.S., and Ingram, B.T. (1983) *Adsorption from Solution at the Solid–Liquid Interface* (eds G.D. Parfitt and C.H. Rochester), Academic Press, London, p. 105.
14. Klimenko, N.A. Tryasorukova A.A., and Permilouskayan, A.A. (1974) *Kolloid. Zh.*, **36**, 678.
15. Kleminko, N.A. (1978) *Kolloid. Zh.*, **40**, 1105.
16. Kleminko, N.A. (1979) *Kolloid. Zh.*, **41**, 781.
17. Corkill, J.M., Goodman, J.F., and Tate, J.R. (1966) *Trans. Faraday Soc.*, **62**, 979.

7
Adsorption and Conformation of Polymeric Surfactants at the Solid–Liquid Interface

Understanding the adsorption and conformation of polymeric surfactants at interfaces is key to know how these molecules act as stabilizers. Most basic ideas on adsorption and conformation of polymers have been developed for the solid–liquid interface [1]. The first theories on polymer adsorption were developed in the 1950s and the 1960s, with extensive developments in the 1970s. The process of polymer adsorption is fairly complicated. In addition to the usual adsorption considerations such as polymer/surface, polymer/solvent, and surface/solvent interactions, one of the principal problems to be resolved is the configuration (conformation) of the polymer at the solid–liquid (S–L) interface. This was recognized by Jenkel and Rumbach in 1951 [2] who found that the amount of polymer adsorbed per unit area of the surface would correspond to a layer more than 10 molecules thick if all the segments of the chain are attached. They suggested a model in which each polymer molecule is attached in sequences separated by bridges that extend into a solution. In other words not all the segments of a macromolecule are in contact with the surface. The segments that are in direct contact with the surface are termed "trains"; those in between and extended into solution are termed "loops"; the free ends of the macromolecule also extending in solution are termed "tails." This is illustrated in Figure 7.1a for a homopolymer. Examples of homopolymers that are formed from the same repeating units are poly(ethylene oxide) or poly(vinyl pyrrolidone). Such homopolymers may adsorb significantly at the S–L interface. Even if the adsorption energy per monomer segment to the surface is small (fraction of kT, where k is the Boltzmann constant and T is the absolute temperature), the total adsorption energy per molecule may be sufficient to overcome the unfavorable entropy loss of the molecule at the S–L interface. Clearly, homopolymers are not the most suitable emulsifiers or dispersants. A small variant is to use polymers that contain specific groups that have high affinity to the surface. This is exemplified by partially hydrolyzed poly(vinyl acetate) (PVAc), technically referred to as poly(vinyl alcohol) (PVA). The polymer is prepared by partial hydrolysis of PVAc, leaving some residual vinyl acetate groups. Most commercially available PVA molecules contain 4–12% acetate groups. These acetate groups that are hydrophobic give the molecule its amphipathic character. On a hydrophobic surface such as polystyrene, the polymer adsorbs with preferential attachment of the acetate groups on the surface, leaving the more hydrophilic vinyl alcohol

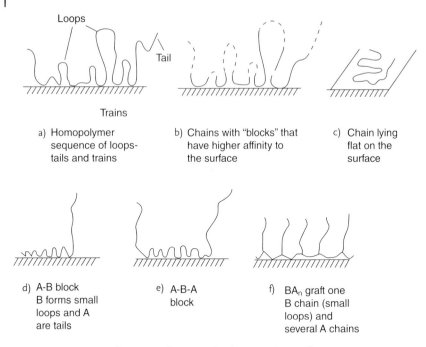

Figure 7.1 Various confirmations of macromolecules on a plane surface.

segments dangling in the aqueous medium. The configuration of such "blocky" copolymers is illustrated in Figure 7.1b. Clearly if the molecule is made fully from hydrophobic segments, the chain will adopt a flat configuration as is illustrated in Figure 7.1c. The most convenient polymeric surfactants are those of the block and graft copolymer type. A block copolymer is a linear arrangement of blocks of variable monomer composition. The nomenclature for a diblock is poly-A-block-poly-B and for a triblock is poly-A-block-poly-B-poly-A. An example of an A–B diblock is polystyrene block-polyethylene oxide and its conformation is represented in Figure 7.1d. One of the most widely used triblock polymeric surfactants are the "Pluronics" (BASF, Germany) that consists of two poly-A blocks of poly(ethylene oxide) (PEO) and one block of poly(propylene oxide) (PPO). Several chain lengths of PEO and PPO are available. As mentioned in Chapter 6, these polymeric triblocks can be applied as dispersants, whereby the assumption is made that the hydrophobic PPO chain resides at the hydrophobic surface, leaving the two PEO chains dangling in the aqueous solution and hence providing steric repulsion. As mentioned in Chapter 6, several other triblock copolymers have been synthesized, although these are of limited commercial availability. Typical examples are triblocks of poly(methyl methacrylate)-block poly(ethylene oxide)-block poly(methyl methacrylate). The conformation of these triblock copolymers is illustrated in Figure 7.1e. An alternative (and perhaps more efficient) polymeric surfactant is the amphipathic graft copolymer consisting of a polymeric backbone B (polystyrene

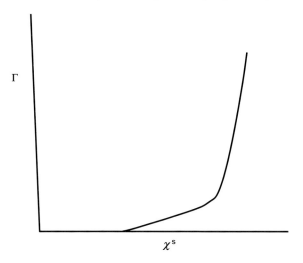

Figure 7.2 Variation of adsorption amount Γ with adsorption energy per segment χ^s.

or polymethyl methacrylate) and several A chains ("teeth") such as polyethylene oxide. This graft copolymer is sometimes referred to as a "comb" stabilizer. Its configuration is illustrated in Figure 7.1f.

The polymer–surface interaction is described in terms of adsorption energy per segment χ^s. The polymer–solvent interaction is described in terms of the Flory–Huggins interaction parameter χ. For adsorption to occur, a minimum energy of adsorption per segment χ^s is required. When a polymer molecule adsorbs on a surface, it looses configurational entropy and this must be compensated by an adsorption energy χ^s per segment. This is schematically shown in Figure 7.2, where the adsorbed amount Γ is plotted versus χ^s. The minimum value of χ^s can be very small (<$0.1\,kT$) since a large number of segments per molecule are adsorbed. For a polymer with say 100 segments and 10% of these are in trains, the adsorption energy per molecule now reaches $1\,kT$ (with $\chi^s = 0.1\,kT$). For 1000 segments, the adsorption energy per molecule is now $10\,kT$.

As mentioned above homopolymers are not the most suitable for stabilization of dispersions. For strong adsorption, one needs the molecule to be "insoluble" in the medium and has strong affinity ("anchoring") to the surface. For stabilization, one needs the molecule to be highly soluble in the medium and strongly solvated by its molecules; this requires a Flory–Huggins interaction parameter less than 0.5. The above opposing effects can be resolved by introducing "short" blocks in the molecule that are insoluble in the medium and have a strong affinity to the surface, as for example partially hydrolyzed polyvinyl acetate (88% hydrolyzed, i.e., with 12% acetate groups), usually referred to as polyvinyl alcohol (PVA),

$$-(CH_2-CH\,)x-(CH_2-CH-)y-(CH_2-CH)x-$$
$$|||$$
$$OHOCOCH_3OH$$

As mentioned earlier these requirements are better satisfied using A–B, A–B–A and BA$_n$ graft copolymers. B is chosen to be highly insoluble in the medium and it should have high affinity to the surface. This is essential to ensure strong "anchoring" to the surface (irreversible adsorption). A is chosen to be highly soluble in the medium and strongly solvated by its molecules. The Flory–Huggins χ parameter can be applied in this case. For a polymer in a good solvent χ has to be lower than 0.5; the smaller the χ value the better the solvent for the polymer chains. Examples of B for hydrophobic particles in aqueous media are polystyrene and polymethylmethacrylate. Examples of A in aqueous media are polyethylene oxide, polyacrylic acid, polyvinyl pyrollidone, and polysaccharides. For nonaqueous media such as hydrocarbons, the A chain(s) could be poly(12-hydroxystearic acid).

For full description of polymer adsorption one needs to obtain information on the following. (i) The amount of polymer adsorbed Γ (in mg or moles) per unit area of the particles. It is essential to know the surface area of the particles in the suspension. Nitrogen adsorption on the powder surface may give such information (by application of the BET equation) provided there will be no change in area on dispersing the particles in the medium. For many practical systems, a change in surface area may occur on dispersing the powder, in which case one has to use dye adsorption to measure the surface area (some assumptions have to be made in this case). (ii) The fraction of segments in direct contact with the surface, that is, the fraction of segments in trains p (p = (Number of segments in direct contact with the surface)/total number). (iii) The distribution of segments in loops and tails, $\rho(z)$, which extend in several layers from the surface. $\rho(z)$ is usually difficult to obtain experimentally although recently application of small angle neutron scattering could obtain such information. An alternative and useful parameter for assessing "steric stabilization" is the hydrodynamic thickness, δ_h (thickness of the adsorbed or grafted polymer layer plus any contribution from the hydration layer). Several methods can be applied to measure δ_h as will be discussed below.

7.1
Theories of Polymer Adsorption

Two main approaches have been developed to treat the problem of polymer adsorption. (i) The random walk approach. This is based on Flory's treatment of the polymer chain in solution; the surface was considered as a reflecting barrier. (ii) The statistical mechanical approach. The polymer configuration was treated as being made of three types of structures, trains, loops, and tails, each having a different energy state.

The random walk approach is based on the random-walk concept that was originally applied to the problem of diffusion and later adopted by Flory [3] to deduce the conformations of macromolecules in solution. The earliest analysis was by Simha et al. [4] who neglected excluded volume effects and treated the polymer as a random walk. Basically the solution was represented by a three-dimensional lattice and the surface by a two-dimensional lattice. The polymer was represented by a realization of a random walk on the lattice. The probabilities of performing steps in different directions were considered to be the same except at the interface that acts as a reflecting barrier. The polymer molecules were, therefore, effectively assumed to be adsorbed with large loops protruding into the solvent and with few segments actually attached to the surface, unless the segment-surface attractive forces were very high. This theory predicts an isotherm for flexible macromolecules that is considerable different from the Langmuir-type isotherm. The number of attached segments per chain is proportional to $n^{1/2}$, where n is the total number of segments. Increasing the molecular weight results in increased adsorption, except for strong chain interaction with the surface.

This approach has been criticized by Silberberg [5] and Di Marzio [6]. One of the major problem was the use of a reflecting barrier as the boundary condition, which meant over counting the number of distinguishable conformations. To overcome this problem, Di Marzio and McCrackin [7] used a Monte Carlo method to calculate the average number of contacts of the chain with the surface, the end-to-end length and distribution of segments $\rho(z)$ with respect to the distance z from the surface, as a function of chain length of the polymer and the attractive energy of the surface. The same method was also used by Clayfield and Lumb [8, 9].

The statistical mechanical approach is a more realistic model for the problem of polymer adsorption since it takes into account the various interactions involved. This approach was first used by Silberberg [5] who treated separately the surface layer that contains adsorbed units (trains) and the adjacent layer in the solution (loops or tails). The units in each layer were considered to be in two different energy states and partition functions were used to describe the system. The units close to the surface are adsorbed with an internal partition function determined by the short-range forces between the segments and the surface, whereas the units in loops and tails were considered to have an internal partition function equivalent to the segments in the bulk. By equating the chemical potential of macromolecules in the adsorbed state and in bulk solution, the adsorption isotherm could be determined. In this treatment, Silberberg [5] assumed a narrow distribution of loop sizes and predicted small loops for all values of the adsorption energy. Later, the loop-size distribution was introduced by Hoeve *et al.* [10] and this theory predicted large loops for small adsorption free energies and small loops and more units adsorbed for larger adsorption free energies when the chains are sufficiently flexible. Most of these theories considered the case of an isolated polymer molecule at an interface, that is, under conditions of low surface coverage, θ. These theories were extended by Silberberg [11] and Hoeve [12, 13] to take into account the lateral interaction between the molecules on the surface, that is, high surface coverage. These theories also considered the excluded volume effect, which reduces the number of configurations available for interacting chains near the surface. Excluded volume effects are strongly dependent on the solvent as is the case for chains in the solution. Some progress has been made in the analysis of the problem of multilayer adsorption [14].

One feature of an adsorbed layer that is important in the theory of steric stabilization is the actual segment distribution normal to the interface. Hoeve [15, 16]

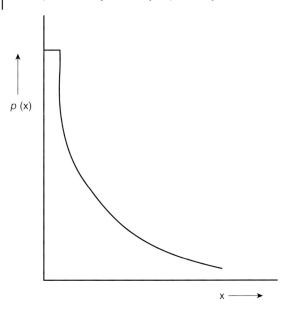

Figure 7.3 Segment density–distance distribution.

was the first to calculate this quantity for an adsorbed homopolymer of loops and tails, using random flight statistics. He showed that at a distance from the interface corresponding to the thickness of the trains, there was a discontinuity in the distribution. Beyond this the segment density falls exponentially with distance, as shown schematically in Figure 7.3. Similarly, Meier [17] developed an equation for the segment density distribution of a single terminally adsorbed tail. Hesselink [18, 19] has developed Meier's theory and given the segment density distribution for single tails, single loops, homopolymers, and random copolymers.

A useful model for treating polymer adsorption and configuration was suggested by Roe [20] Scheutjens and Fleer (SF theory) [21–24] that is referred to as the step weighted random walk approach. In order to be able to describe all possible chain conformations, Scheutjens and Fleer [20–22] used a model of a quasicrystalline lattice with lattice layers parallel to the surface. Starting from the surface the layers are numbered I = 1, 2, 3,..., M where M is a layer in bulk solution. All the lattice sites within one layer were considered to be energetically equivalent. The probability of finding any lattice site in layer I occupied by a segment was assumed to be equal to the volume fraction ϕ_I in this layer. The conformation probability and the free energy of mixing were calculated with the assumption of random mixing within each layer (the Brag–Williams or mean-field approximation). The energy for any segment is only determined by the layer number, and each segment can be assigned a weighting or Boltzmann factor p_i that depends only on the layer number. The partition functions were derived for the mixture of free and adsorbed polymer molecules, as well as for the solvent molecules. As mentioned earlier, all

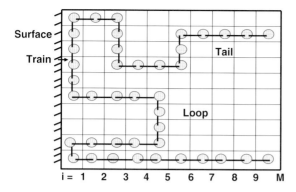

Figure 7.4 Schematic representation of a polymer molecule adsorbing on a flat surface–quasicrystalline lattice with segments filling layers that are parallel to the surface (random mixing of segments and solvent molecules in each layer is assumed).

chain conformations were described as step weighted random walks on a quasicrystalline lattice that extends in parallel layers from the surface; this is schematically shown in Figure 7.4. The partition function is written in terms of a number of configurations. These were treated as connected sequences of segments. In each layer, random mixing between segments and solvent molecules was assumed (mean-field approximation). Each step in the random walk was assigned a weighting factor p_i that consists of three contributions. (i) An adsorption energy χ^s (that exists only for the segments that are near the surface). (ii) Configurational entropy of mixing (that exists in each layer). (iii) Segment–solvent interaction parameter χ (the Flory–Huggins interaction parameter; note that $\chi = 0$ for an athermal solvent; $\chi = 0.5$ for a θ-solvent). The adsorption energy gives rise to a Boltzmann factor exp χ_s in the weighting factor for the first layer, provided χ_s is interpreted as the adsorption energy difference (in units of kT) between a segment and a solvent molecule. The configurational entropy for the segment, as a part of the chain, is accounted for in the matrix procedure in which all possible chain conformations are considered. However, the configurational entropy loss of the solvent molecule, going from a layer i with low solvent concentration to the bulk solution with a higher solvent concentration, has to be introduced in p_i. According to the Flory–Huggins theory [3], this entropy loss can be written as $\Delta s^0 = k \ln \phi_*^0 / \phi_i^0$ per solvent molecule, where ϕ_i^0 and ϕ_*^0 are the solvent volume fractions in layer I and in bulk solution, respectively. This change is equivalent to introducing a Boltzmann factor exp $(-\Delta s^0 / k) = \phi_i^0 / \phi_*^0$ in the weighting factor p_i. The last contribution stems from the mixing energy of the exchange process. The transfer of a segment from the bulk solution to layer i is accompanied by an energy change (in units of kT) $\chi(\phi_i^0 - \phi_*^0)$, where χ is the Flory–Huggins segment solvent interaction parameter.

Figure 7.5 shows typical adsorption isotherms plotted as surface coverage (in equivalent monolayers) versus polymer volume fraction ϕ_* in the bulk solution (ϕ_* was taken to vary between 0 and 10^{-3} that is the normal experimental range). The results in Figure 7.6 show the effect of increasing the chain length r and effect of

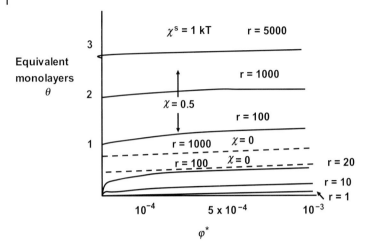

Figure 7.5 Adsorption isotherms for oligomers and polymers in the dilute region based on the SF theory. Full curves $\chi = 0.5$; dashed curves $\chi = 0$.

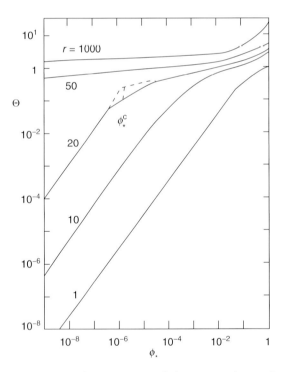

Figure 7.6 Log–log presentation of adsorption isotherms of various r values, $\chi_s = 1$; $\chi = 0.5$. Hexagonal lattice.

solvency (athermal solvent with $\chi = 0$ and theta solvent with $\chi = 0.5$). The adsorption energy χ^s was taken to be the same and equal to $1\,kT$. When $r = 1$, θ is very small and the adsorption increases linearly with increase of ϕ^* (Henry's type isotherm). On the other hand when $r = 10$, the isotherm deviates much from a straight line and approaches a Langmuirian type. However, when $r \geq 20$ high affinity isotherms are obtained. This implies that the first added polymer chains are completely adsorbed resulting in extremely low polymer concentration in solution (approaching zero). This explains the irreversibility of adsorption of polymeric surfactants with $r > 100$. The adsorption isotherms with $r = 100$ and above are typical of those observed experimentally for most polymers that are not too polydisperse, that is, showing a steep rise followed by a nearly horizontal plateau (that only increases few percent per decade increase of ϕ^*). In these dilute solutions, the effect of solvency is most clearly seen, with poor solvents giving the highest adsorbed amounts. In good solvents θ is much smaller and levels off for long chains to attain an adsorption plateau that is essentially independent of molecular weight.

Some general features of the adsorption isotherms over a wide concentration range can be illustrated by using logarithmic scales for both θ and φ_* which highlight the behavior in extremely dilute solutions. Such presentation [23] is shown in Figure 7.6. These results show a linear Henry region followed by a pseudoplateau region. A transition concentration, φ_*^{1c}, can be defined by extrapolation of the two linear parts. φ_*^c decreases exponentially with increasing chain length and when $r = 50$, φ_*^c is so small (10^{-12}) that it does not appear within the scale shown in Figure 7.6. With $r = 1000$, φ_*^c reaches the ridiculously low value of 10^{-235}. The region below φ_*^c is the Henry region where the adsorbed polymer molecules behaves essentially as isolated molecules. The representation in Figure 7.7 also answers the question of reversibility versus irreversibility for polymer adsorption. When $r > 50$, the pseudoplateau region extends down to very low concentration ($\varphi_*^c = 10^{-12}$) and this explains why one cannot easily detect any desorption upon dilution. Clearly if such extremely low concentration can be reached, desorption of the polymer may take place. Thus, the lack of desorption (sometimes referred to as irreversible adsorption) is due to the fact that the equilibrium between adsorbed and free polymer is shifted far in favor of the surface because of the high possible number of possible attachments per chain.

Another point that emerges from the SF theory is the difference in the shape between the experimental and theoretical adsorption isotherms in the low concentration region. The experimental isotherms are usually rounded, whereas those predicted from theory are flat. This is accounted for in terms of the molecular weight distribution (polydispersity) that is encountered with most practical polymers. This effect has been explained by Cohen-Stuart and Mulder [25]. With polydisperse polymers, the larger molecular weight fractions adsorb preferentially over the smaller ones. At low polymer concentrations, nearly all polymer fractions are adsorbed leaving a small fraction of the polymer with the lowest molecular weights in solution. As the polymer concentration is increased, the higher molecular weight fractions displace the lower ones on the surface, which are now released

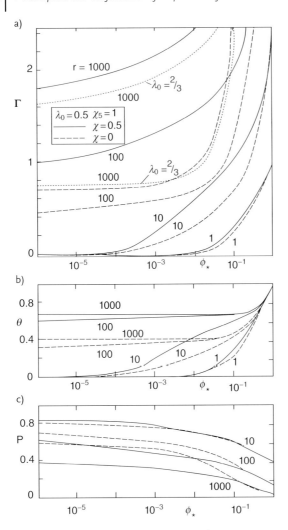

Figure 7.7 (a) Adsorbed amount Γ, (b) surface coverage θ, (c) and fraction of adsorbed segments $p = \theta/\Gamma$ as a function of volume fraction ϕ_*. Full lines for a θ-solvent ($\chi = 0.5$), broken line for an athermal solvent ($\chi = 0$).

in solution, thus shifting the molecular weight distribution in solution to lower values. This process continues with further increase in polymer concentration leading to fractionation whereby the higher molecular weight fractions are adsorbed at the expense of the lower fractions that are released to the bulk solution. However, in very concentrated solutions, monomers adsorb preferentially with respect to polymers and short chains with respect to larger ones. This is due to the fact that in this region, the conformational entropy term predominates the free energy, disfavoring the adsorption of long chains.

According to the SF theory, the bound fraction p and the direct surface coverage θ_1 depends on the chain length for the same volume fraction. This is illustrated in Figure 7.7 that shows the adsorbed amount Γ (Figure 7.7a), surface coverage θ (Figure 7.7b), and fraction of adsorbed segments $p = \theta/\Gamma$ (Figure 7.7c) as a function of volume fraction ϕ_*.

In the Henry region ($\phi_* < \phi_*^c$), p is rather high and independent of chain length for $r \geq 20$. In this region, the molecules lie nearly flat on the surface, with 87% of segments in trains. At the other end of the concentration range ($\phi_* = 1$), p is proportional to $r^{-1/2}$. At intermediate concentrations, p is within these two extremes. With increasing polymer concentration, the adsorbed molecules become gradually more extended (lower p) until at very high ϕ_* values they become Gaussian at the interface. In better solvents the direct surface coverage is lower due to the stronger repulsion between the segments. This effect is more pronounced if the surface concentration differs strongly from the solution concentration. If the adsorption is small, the effect of the excluded volume effect (and therefore of χ) is rather weak; the same applies if both the concentrations in the bulk solution and near the surface are high. Both θ_1 and θ decrease with increasing solvent power (decreasing χ) but the effect is stronger for θ than for θ_1, resulting in a higher bound fraction (thus flatter chains) from better solvents at the same solution concentration.

The structure of the adsorbed layer is described in terms of the segment density distribution. As an illustration, Figure 7.8 shows some calculations using the SF theory for loops and tails with $r = 1000$, $\phi^* = 10^{-6}$ and $\chi = 0.5$. In this example, 38% of the segments are in trains, 55.5% in loops, and 6.5% in tails. This theory demonstrates the importance of tails that dominate the total distribution in the outer region.

7.2
Experimental Techniques for Studying Polymeric Surfactant Adsorption

As mentioned above, for full characterization of polymeric surfactant adsorption one needs to determine three parameters. (i) The adsorbed amount Γ (mg m^{-2} or mol m^{-2}) as a function of equilibrium concentration C_{eq}, that is, the adsorption

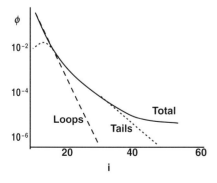

Figure 7.8 Loop, tail and total segment profile according to the SF theory.

isotherm. (ii) The fraction of segments in direct contact with the surface p (number of segments in trains relative to the total number of segments). (iii) The segment density distribution $\rho(z)$ or the hydrodynamic adsorbed layer thickness δ_h.

It is important to obtain the adsorption parameters as a function of the important variables of the system: (i) Solvency of the medium for the chain that can be affected by temperature, addition of salt or a nonsolvent. The Flory–Huggins interaction parameter χ could be separately measured. (ii) The molecular weight of the adsorbed polymer. (iii) The affinity of the polymer to the surface as measured by the value of χ^s, the segment-surface adsorption energy. (iv) The structure of the polymer; this is particularly important for block and graft copolymers.

7.3
Measurement of the Adsorption Isotherm

This is by far the easiest to obtain. One measures the polymeric surfactant concentration before ($C_{initial}$, C_1) and after ($C_{equilibrium}$, C_2)

$$\Gamma = \frac{(C_1 - C_2)V}{A} \tag{7.1}$$

where V is the total volume of the solution and A is the specific surface area (m^2 g^{-1}). It is necessary in this case to separate the particles from the polymer solution after adsorption. This could be carried out by centrifugation and/or filtration. One should make sure that all particles are removed. To obtain this isotherm, one must develop a sensitive analytical technique for determination of the polymeric surfactant concentration in the ppm range. It is essential to follow the adsorption as a function of time to determine the time required to reach equilibrium. For some polymer molecules such as polyvinyl alcohol, PVA, and polyethylene oxide, PEO, (or blocks containing PEO), analytical methods based on complexation with iodine–potassium iodide or iodine–boric acid potassium iodide have been established. For some polymers with specific functional groups spectroscopic methods may be applied, for example, UV, IR, or fluorescence spectroscopy. A possible method is to measure the change in refractive index of the polymer solution before and after adsorption. This requires very sensitive refractometers. High-resolution NMR has been recently applied since the polymer molecules in the adsorbed state are in different environment than those in the bulk. The chemical shift of functional groups within the chain is different in these two environments. This has the attraction of measuring the amount of adsorption without separating the particles.

7.4
Measurement of the Fraction of Segments p

The fraction of segments in direct contact with the surface can be directly measured using spectroscopic techniques: (i) IR if there is specific interaction between the segments in trains and the surface, for example, polyethylene oxide on silica from nonaqueous solutions [26, 27]. (ii) Electron spin resonance (ESR); this requires labeling of the molecule. (iii) NMR, pulse gradient or spin ECO NMR. This method is based on the fact that the segments in trains are "immobilized" and hence they have lower mobility than those in loops and tails [28, 29].

An indirect method of determination of p is to measure the heat of adsorption ΔH using microcalorimetry [30]. One should then determine the heat of adsorption of a monomer H_m (or molecule representing the monomer, for example, ethylene glycol for PEO); p is then given by the equation

$$p = \frac{\Delta H}{H_m n} \tag{7.2}$$

where n is the total number of segments in the molecule.

The above indirect method is not very accurate and can only be used in a qualitative sense. It also requires very sensitive enthalpy measurements (e.g., using an LKB microcalorimeter).

7.5
Determination of the Segment Density Distribution $\rho(z)$ and Adsorbed Layer Thickness δ_h

The segment density distribution $\rho(z)$ is given by the number of segments parallel to the surface in the z-direction. Three direct methods can be applied for determination of adsorbed layer thickness: ellipsometry, attenuated total reflection (ATR), and neutron scattering. Both ellipsometry and ATR [31] depend on the difference between refractive indices between the substrate, the adsorbed layer, and bulk solution and they require flat reflecting surface. Ellipsometry [31] is based on the principle that light undergoes a change in polarizability when it is reflected at a flat surface (whether covered or uncovered with a polymer layer.

The above limitations when using ellipsometry or ATR are overcome by the application technique of neutron scattering, which can be applied to both flat surfaces as well as particulate dispersions. The basic principle of neutron scattering is to measure the scattering due to the adsorbed layer, when the scattering length density of the particle is matched to that of the medium (the so-called contrast-matching method). Contrast matching of particles and medium can be achieved by changing the isotopic composition of the system (using deuterated particles and mixture of D_2O and H_2O). It was used for measurement of the adsorbed layer thickness of polymers, for example, PVA or poly(ethylene oxide) (PEO) on polystyrene latex [32]. Apart from obtaining δ, one can also determine the segment density distribution $\rho(z)$.

The above technique of neutron scattering gives clearly a quantitative picture of the adsorbed polymer layer. However, its application in practice is limited s ince one need to prepare deuterated particles or polymers for the contrast matching procedure. The practical methods for determination of the adsorbed layer thickness are mostly based on hydrodynamic methods. Several methods may be applied to determine the hydrodynamic thickness of adsorbed polymer layers of which viscosity, sedimentation coefficient (using an ultracentrifuge), and dynamic light scattering measurements are the most convenient. A less accurate method is from zeta potential measurements.

The viscosity method [33] depends on measuring the increase in the volume fraction of the particles as a result of the presence of an adsorbed layer of thickness δ_h. The volume fraction of the particles φ plus the contribution of the adsorbed layers is usually referred to as the effective volume fraction φ_{eff}. Assuming the particles behave as hard spheres, then the measured relative viscosity η_r is related to the effective volume fraction by Einstein's equation, that is,

$$\eta_r = 1 + 2.5\varphi_{\text{eff}} \tag{7.3}$$

φ_{eff} and φ are related from simple geometry by

$$\varphi_{\text{eff}} = \varphi\left[1 + \left(\frac{\delta_h}{R}\right)\right]^3 \tag{7.4}$$

where R is the particle radius. Thus, from a knowledge of η_r and φ one can obtain δ_h using the above equations.

The sedimentation method depends on measuring the sedimentation coefficient (using an ultracentrifuge) of the particles S'_0 (extrapolated to zero concentration) in the presence of the polymer layer [34]. Assuming the particles obey Stokes' law, S'_0 is given by the expression

$$S_{0'} = \frac{(4/3)\pi R^3(\rho - \rho_s) + (4/3)\pi[(R+\delta_h)^3 - R^3](\rho_s^{\text{ads}} - \rho_s)}{6\pi\eta(R+\delta_h)} \tag{7.5}$$

where ρ and ρ_s are the mass density of the solid and solution phase, respectively, and ρ^{ads} is the average mass density of the adsorbed layer that may be obtained from the average mass concentration of the polymer in the adsorbed layer.

In order to apply the above-stated methods, one should use a dispersion with monodisperse particles with a radius that is not much larger than δ_h. Small model particles of polystyrene may be used.

A relatively simple sedimentation method for determination of δ_h is the slow speed centrifugation applied by Garvey et al. [34]. Basically a stable monodisperse dispersion is slowly centrifuged at low g values (<50g) to form a close-packed (hexagonal or cubic) lattice in the sediment. From a knowledge of φ and the packing fraction (0.74 for hexagonal packing), the distance of separation between the center of two particles R_δ may be obtained, that is,

$$R_\delta = R + \delta_h = \left(\frac{0.74 V \rho_1 R^3}{W}\right) \tag{7.6}$$

where V is the sediment volume, ρ_1 is the density of the particles and W their weight.

The most rapid technique for measuring δ_h is photon correlation spectroscopy (PCS) (sometime referred to as quasielastic light scattering) that allows one to

7.5 Determination of the Segment Density Distribution ρ(z) and Adsorbed Layer Thickness δ_h

obtain the diffusion coefficient of the particles with and without the adsorbed layer (D_δ and D, respectively). This is obtained from measurement of the intensity fluctuation of scattered light as the particles undergo Brownian diffusion [35]. When a light beam (e.g., monochromatic laser beam) passes through a dispersion, an oscillating dipole is induced in the particles, thus re-radiating the light. Due to the random arrangement of the particles (that are separated by a distance comparable to the wavelength of the light beam, that is, the light is coherent with the interparticle distance), the intensity of the scattered light will, at any instant, appear as random diffraction or "speckle" pattern. As the particles undergo the Brownian motion, the random configuration of the speckle pattern changes. The intensity at any one point in the pattern will, therefore, fluctuate such that the time taken for an intensity maximum to become a minimum (i.e., the coherence time) corresponds approximately to the time required for a particle to move one wavelength. Using a photomultiplier of active area about the size of a diffraction maximum, that is, approximately one coherence area, this intensity fluctuation can be measured. A digital correlator is used to measure the photocount or intensity correlation function of the scattered light. The photocount correlation function can be used to obtain the diffusion coefficient D of the particles. For monodisperse noninteracting particles (i.e., at sufficient dilution), the normalized correlation function $[g^{(1)}(\tau)]$ of the scattered electric field is given by the equation

$$[g^{(1)}(\tau)] = \exp-(\Gamma\tau) \tag{7.7}$$

where τ is the correlation delay time and Γ is the decay rate or inverse coherence time. Γ is related to D by the equation

$$\Gamma = DK^2 \tag{7.8}$$

where K is the magnitude of the scattering vector that is given by

$$K = \left(\frac{4n}{\lambda_0}\right)\sin\left(\frac{\theta}{2}\right) \tag{7.9}$$

where n is the refractive index of the solution, λ is the wavelength of light in vacuum and θ is the scattering angle.

From D, the particle radius R is calculated using the Stokes–Einstein equation:

$$D = \frac{kT}{6\pi\eta R} \tag{7.10}$$

where k is the Boltzmann constant and T is the absolute temperature. For a polymer-coated particle R is denoted by R_δ that is equal to $R + \delta_h$. Thus, by measuring D_δ and D, one can obtain δ_h. It should be mentioned that the accuracy of the PCS method depends on the ratio of δ_δ/R, since δ_h is determined by difference. Since the accuracy of the measurement is ±1%, δ_h should be at least 10% of the particle radius. This method can only be used with small particles and reasonably thick adsorbed layers. Electrophoretic mobility, u, measurements can also be applied to measure δ_h [36]. From u, the zeta potential ζ, that is, the potential at the slipping (shear) plane of the particles can be calculated. Adsorption of a polymer causes a shift in the shear plane from its value in the absence of a polymer layer (that is close to the Stern plane) to a value that depends on the thickness of the adsorbed layer. Thus, by measuring ζ in the presence (ζ_δ) and absence (ζ) of a polymer layer one can estimate δ_h. Assuming that the thickness of the Stern plane is Δ, then ζ_δ may be related to the ζ (that may be assumed to be equal to the Stern potential ψ_d) by the equation

$$\tanh\left(\frac{e\psi_\delta}{4\,\mathrm{kT}}\right) = \tanh\left(\frac{e\zeta}{4\,\mathrm{kT}}\right)\exp(-\kappa(\delta_h - \Delta)) \tag{7.11}$$

where κ is the Debye parameter that is related to electrolyte concentration and valency.

It should be mentioned that the value of δ_h calculated using the above simple equation shows a dependence on electrolyte concentration and hence the method cannot be used in a straightforward manner. Cohen-Stuart and Mulder [36] showed that the measured electrophoretic thickness δ_e approaches δ_h only at low electrolyte concentrations. Thus, to obtain δ_h from electrophoretic mobility measurements, results should be obtained at various electrolyte concentrations and δ_e should be plotted versus the Debye length $(1/\kappa)$ to obtain the limiting value at high $(1/\kappa)$ (i.e., low electrolyte concentration) that now corresponds to δ_h.

7.6
Examples of the Adsorption Isotherms of Nonionic Polymeric Surfactants

Figure 7.9 shows the adsorption isotherms for PEO with different molecular weights on PS (at room temperature). It can be seen that the amount adsorbed in mg m^{-2} increases with increase in the polymer molecular weight [37]. Figure 7.10 shows the variation of the hydrodynamic thickness δ_h with molecular weight M.

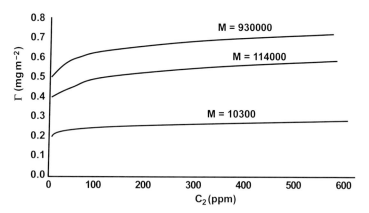

Figure 7.9 Adsorption isotherms for PEO on PS.

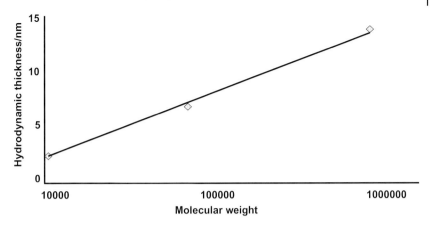

Figure 7.10 Hydrodynamic thickness of PEO on PS as a function of the molecular weight.

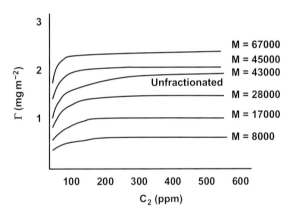

Figure 7.11 Adsorption isotherms of PVA with different molecular weights on polystyrene latex at 25 °C.

δ_h shows a linear increase with log M. δ_h increases with n, the number of segments in the chain according to

$$\delta_h \approx n^{0.8} \tag{7.12}$$

Figure 7.11 shows the adsorption isotherms of PVA with various molecular weights on PS latex (at 25 °C) [38]. The polymers were obtained by fractionation of a commercial sample of PVA with an average molecular weight of 45 000. The polymer also contained 12% vinyl acetate groups. As with PEO, the amount of adsorption increases with increase in M. The isotherms are also of the high affinity type. Γ at the plateau increases linearly with $M^{1/2}$.

The hydrodynamic thickness was determined using PCS and the results are given below

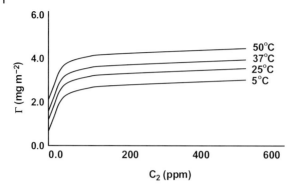

Figure 7.12 Influence of temperature on adsorption.

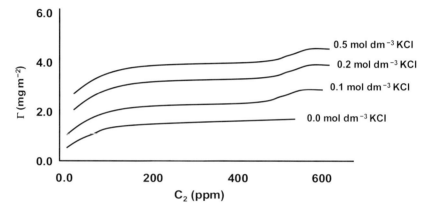

Figure 7.13 Influence of addition of KCl on adsorption.

M	67 000	43 000	28 000	17 000	8000
δ_h (nm)	25.5	19.7	14.0	9.8	3.3

δ_h seems to increase linearly with increase in the molecular weight.

The effect of solvency on adsorption was investigated by increasing the temperature (the PVA molecules are less soluble at higher temperature) or addition of electrolyte (KCl) [39, 40]. The results are shown in Figures 7.12 and 7.13 for $M = 65\,100$. As can be seen from Figure 7.12 increase of temperature results in reduction of solvency of the medium for the chain (due to break down of hydrogen bonds) and this results in an increase in the amount adsorbed. Addition of KCl (that reduces the solvency of the medium for the chain) results in an increase in adsorption (as predicted by theory).

The adsorption of block and graft copolymers is more complex since the intimate structure of the chain determines the extent of adsorption [37]. Random

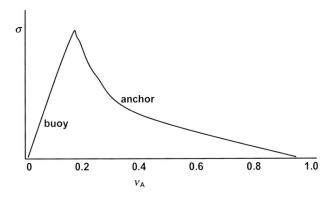

Figure 7.14 Prediction of adsorption of diblock copolymer.

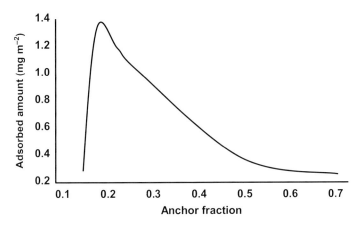

Figure 7.15 Adsorbed amount (mg m^{-2}) versus fraction of anchor segment for an A–B–A triblock copolymer (PEO-PPO-PEO).

copolymers adsorb in an intermediate way to that of the corresponding homopolymers. Block copolymers retain the adsorption preference of the individual blocks. The hydrophilic block (e.g., PEO), the buoy, extends away from the particle surface into the bulk solution, whereas the hydrophobic anchor block (e.g., PS or PPO) provides firm attachment to the surface. Figure 7.14 shows the theoretical prediction of diblock copolymer adsorption according to the Scheutjens and Fleer theory. The surface density σ is plotted versus the fraction of anchor segments v_A. The adsorption depends on the anchor/buoy composition.

The amount of adsorption is higher than for homopolymers and the adsorbed layer thickness is more extended and dense. For a triblock copolymer A–B–A, with two buoy chains and one anchor chain, the behavior is similar to that of diblock copolymers. This is shown in Figure 7.15 for PEO-PPO-PEO block (Pluronic).

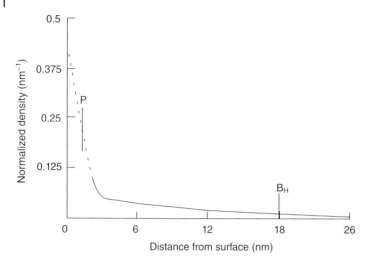

Figure 7.16 Plot of $\rho(z)$ against z for PVA ($M = 37\,000$) adsorbed on deuterated PS latex in D_2O/H_2O.

7.7
Adsorbed Layer Thickness Results

Figure 7.16 shows a plot of $\rho(z)$ against z for PVA ($M = 37\,000$) adsorbed on deuterated PS latex in D_2O/H_2O.

The results shows a monotonic decay of $\rho(z)$ with distance z from the surface and several regions may be distinguished. Close to the surface ($0 < z < 3$ nm), the decay in $\rho(z)$ is rapid and assuming a thickness of 1.3 nm for the bound layer, p was calculated to be 0.1, which is in close agreement with the results obtained using NMR measurements. In the middle region, $\rho(z)$ shows a shallow maximum followed by a slow decay which extends to 18 nm, that is, close to the hydrodynamic layer thickness δ_h of the polymer chain (see below). δ_h is determined by the longest tails and is about 2.5 times the radius of gyration in bulk solution (~7.2 nm). This slow decay of $\rho(z)$ with z at long distances is in qualitative agreement with Scheutjens and Fleers's theory [23] that predicts the presence of long tails. The shallow maximum at intermediate distances suggests that the observed segment density distribution is a summation of a fast monotonic decay due to loops and trains together with the segment density for tails which a maximum density away from the surface. The latter maximum was clearly observed for a sample which had PEO grafted to a deuterated polystyrene latex [32] (where the configuration is represented by tails only.

The hydrodynamic thickness of block copolymers shows different behavior from that of homopolymers (or random copolymers). Figure 7.17 shows the theoretical prediction for the adsorbed layer thickness δ that is plotted as a function of v_A.

Figure 7.18 shows the hydrodynamic thickness versus fraction of anchor segment for an ABA block copolymer of (polyethylene oxide)-poly(propylene

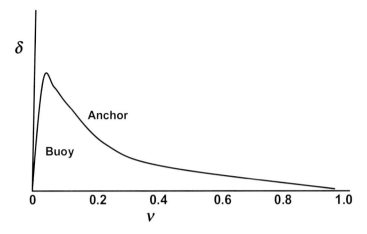

Figure 7.17 Theoretical predictions of the adsorbed layer thickness for a diblock copolymer.

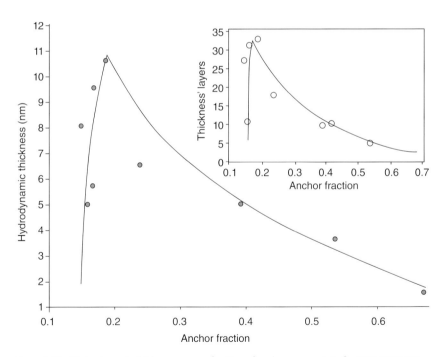

Figure 7.18 Hydrodynamic thickness versus fraction of anchor segment v_A for PEO-PPO-PEO block copolymer onto polystyrene latex. Insert shows the mean field calculation of thickness versus anchor fraction using the SF theory.

oxide)-poly(ethylene oxide) (PEO-PPO-PEO) [37] versus fraction of anchor segment. The theoretical (Scheutjens and Fleer) prediction of adsorbed amount and layer thickness versus fraction of anchor segment are shown in the inserts of Figure 7.18. When there are two buoy blocks and a central anchor block, as in the above example, the A–B–A block shows similar behavior to that of an A–B block. However, if there are two anchor blocks and a central buoy block, surface precipitation of the polymer molecule at the particle surface is observed and this is reflected in a continuous increase of adsorption with increase in polymer concentration as has been show for an A–B–A block of PPO-PEO-PPO [37].

7.8
Kinetics of Polymer Adsorption

The kinetics of polymer adsorption is a highly complex process. Several distinct processes can be distinguished, each with a characteristic time scale [37]. These processes may occur simultaneously and hence it is difficult to separate them. The first process is the mass transfer of the polymer to the surface, which may be either diffusion or convection. Having reached the surface, the polymer must then attach itself to a surface site, which depends on any local activation energy barrier. Finally, the polymer will undergo large-scale rearrangements as it changes from its solution conformation to a "tail-train-loop" conformation. Once the polymer has reached the surface, the amount of adsorption increases with time. The increase is rapid at the beginning but subsequently slows as the surface becomes saturated. The initial rate of adsorption is sensitive to the bulk polymer solution concentration and molecular weight as well as the solution viscosity. Nevertheless, all the polymer molecules arriving at the surface tend to adsorb immediately. The concentration of unadsorbed polymer around the periphery of the forming layer (the surface polymer solution) is zero and, therefore the concentration of polymer in the interfacial region is significantly greater than the bulk polymer concentration. Mass transport is found to dominate the kinetics of adsorption until 75% of full surface coverage. At higher surface coverage, the rate of adsorption decreases since the polymer molecules arriving at the surface cannot immediately adsorb. Overtime equilibrium is set up between this interfacial concentration of polymer and the concentration of polymer in the bulk. Given that the adsorption isotherm is of the high affinity type, no significant change in adsorbed amount is expected, even over decades of polymer concentration. If the surface polymer concentration increases toward that of the bulk solution, the rate of adsorption decreases because the driving force for adsorption (the difference in concentration between the surface and bulk solutions) decreases. Adsorption processes tend to be very rapid and an equilibrated polymer layer can form within several 1000 s. However, desorption is a much slower process and this can take several years.

References

1. Tadros, T.F. (1985) *Polymer Colloids* (eds R. Buscall, T. Corner, and J.F. Stageman), Elsevier Applied Sciences, London, p. 105.
2. Jenkel, E., and Rumbach, R. (1951) *Z. Elektrochem.*, **55**, 612.
3. Flory, P.J. (1953) *Principles of Polymer Chemistry*, Cornell University Press, New York.
4. Simha, R., Frisch, L., and Eirich, F.R. (1953) *J. Phys. Chem.*, **57**, 584.
5. Silberberg, A. (1962) *J. Phys. Chem.*, **66**, 1872.
6. Di Marzio, E.A. (1965) *J. Chem. Phys.*, **42**, 2101.
7. Di Marzio, E.A., and McCrakin, F.L. (1965) *J. Chem. Phys.*, **43**, 539.
8. Clayfield, E.J., and Lumb, E.C. (1966) *J. Colloid Interface Sci.*, **22**, 269, 285.
9. Clayfield, E.J., and Lumb, E.C. (1968) *Macromolecules*, **1**, 133.
10. Hoeve, C.A., Di Marzio, E.A., and Peyser, P. (1965) *J. Chem. Phys.*, **42**, 2558.
11. Silberberg, A. (1968) *J. Chem. Phys.*, **48**, 2835.
12. Hoeve, C.A. (1965) *J. Chem. Phys.*, **44**, 1505; Hoeve, C.A. (1966) *J. Chem. Phys.*, **47**, 3007.
13. Hoeve, C.A. (1970) *J. Polym. Sci.*, **30**, 361; Hoeve, C.A. (1971) *J. Polym. Sci.*, **34**, 1.
14. Silberberg, A. (1972) *J. Colloid Interface Sci.*, **38**, 217.
15. Hoeve, C.A.J. (1965) *J. Chem. Phys.*, **44**, 1505.
16. Hoeve, C.A.J. (1966) *J. Chem. Phys.*, **47**, 3007.
17. Meier, D.J. (1861) *J. Phys. Chem.*, **71**, 1965.
18. Hesselink, F.T. (1969) *J. Phys. Chem.*, **73**, 3488.
19. Hesselink, F.T. (1971) *J. Phys. Chem.*, **75**, 65.
20. Roe, R.J. (1974) *J. Chem. Phys.*, **60**, 4192.
21. Scheutjens, J.M.H.M., and Fleer, G.J. (1979) *J. Phys. Chem.*, **83**, 1919.
22. Scheutjens, J.M.H.M., and Fleer, G.J. (1980) *J. Phys. Chem.*, **84**, 178.
23. Scheutjens, J.M.H.M., and Fleer, G.J. (1982) *Adv. Colloid Interface Sci.*, **16**, 341.
24. Fleer, G.J., Cohen-Stuart, M.A., Scheutjens, J.M.H.M., Cosgrove, T., and Vincent, B. (1993) *Polymers at Interfaces*, Chapman and Hall, London.
25. Cohen-Stuart, M.A., Scheutjens, J.M.H.M., and Fleer, G.J. (1980) *J. Polym. Sci. Polym. Phys. Ed.*, **18**, 559.
26. Killmann, E., Eisenlauer, E., and Horn, M. (1977) *J. Polym. Sci. Polym. Symp.*, **61**, 413.
27. Fontana, B.J., and Thomas, J.R. (1961) *J. Phys. Chem.*, **65**, 480.
28. Robb, I.D., and Smith, R. (1974) *Eur. Polym. J.*, **10**, 1005.
29. Barnett, K.G., Cosgrove, T., Vincent, B., Burgess, A., Crowley, T.L., Kims, J., Turner, J.D., and Tadros, T.F. (1981) *Discuss. Faraday Soc.*, **22**, 283.
30. Cohen-Staurt, M.A., Fleer, G.J., and Bijesterbosch, B. (1982) *J. Colloid Interface Sci.*, **90**, 321.
31. Abeles, F. (1964) *Ellipsometry in the Measurement of Surfaces and Thin Films*, vol. 256 (eds E. Passaglia, R.R. Stromberg, and J. Kruger), p. 41. Nat. Bur. Stand. Misc. Publ, Washington, DC.
32. Cosgrove, T., Crowley, T.L., and Ryan, T. (1987) *Macromolecules*, **20**, 2879.
33. Einstein, A. (1906) *Investigations on the Theory of the Brownian Movement*, Dover, New York.
34. Garvey, M.J., Tadros, T.F., and Vincent, B. (1976) *J. Colloid Interface Sci.*, **55**, 440.
35. Pusey, P.N. (1973) *Industrial Polymers: Characterisation by Molecular Weights*, (eds J.H.S. Green and R. Dietz), Transcripta Books, London.
36. Cohen-Stuart, M.A., and Mulder, J.W. (1985) *Colloids Surf.*, **15**, 49.
37. Obey, T.M., and Griffiths, P.C. (1999) Chapter 2, in *Principles of Polymer Science and Technology in Cosmetics and Personal Care*, (eds E.D. Goddard and J.V. Gruber), Marcel Dekker, New York.
38. Garvey, M.J., Tadros, T.F., and Vincent, B. (1974) *J. Colloid Interface Sci.*, **49**, 57.
39. van den Boomgaard, T., King, T.A., Tadros, T.F., Tang, H., and Vincent, B. (1978) *J. Colloid Interface Sci.*, **61**, 68.
40. Tadros, T.F., and Vincent, B. (1978) *J. Colloid Interface Sci.*, **72**, 505.

8
Stabilization and Destabilization of Suspensions Using Polymeric Surfactants and the Theory of Steric Stabilization

8.1
Introduction

The use of natural and synthetic polymers (referred to as polymeric surfactants) for stabilization and destabilization of solid–liquid dispersions plays an important role in industrial application, such as in paints, cosmetics, agrochemicals, ceramics, etc. Polymers are particularly important for preparation of concentrated dispersions, that is, at high volume fraction φ of the disperse phase:

ϕ = (volume of all particles)/(total volume of dispersion)

Polymers are also essential for the stabilization of nonaqueous dispersions, since in this case electrostatic stabilization is not possible (due to the low dielectric constant of the medium). To understand the role of polymers in dispersion stability, it is essential to consider the adsorption and conformation of the macromolecule at the solid–liquid interface and this was discussed in detail in Chapter 7. Polymers and polyelectrolytes are also used for destabilization of suspensions, for example, for solid–liquid separation. In this chapter, we will cover the following topics. In Section 8.2 the interaction between particles containing adsorbed polymeric surfactants and the theory of steric stabilization will be discussed. In Section 8.3 flocculation of sterically stabilized dispersions: weak, incipient, and depletion flocculation will be discussed. Section 8.4 discusses bridging flocculation by polymers and polyelectrolytes. Section 8.5 discusses the examples for stabilization of suspensions using polymeric surfactants.

8.2
Interaction between Particles Containing Adsorbed Polymeric Surfactant Layers (Steric Stabilization)

When two particles each with a radius R and containing an adsorbed polymer layer with a hydrodynamic thickness δ_h, approach each other to a surface–surface separation distance h that is smaller than $2\delta_h$, the polymer layers interact with each other resulting in two main situations [1]. (i) The polymer chains may overlap with

Dispersion of Powders in Liquids and Stabilization of Suspensions, First Edition. Tharwat F. Tadros.
© 2012 Wiley-VCH Verlag GmbH & Co. KGaA. Published 2012 by Wiley-VCH Verlag GmbH & Co. KGaA.

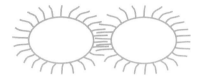
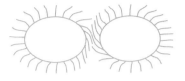

Interpenetration without compression

Compression without interpenetration

Figure 8.1 Schematic representation of the interaction between particles containing adsorbed polymer layers.

each other. (ii) The polymer layer may undergo some compression. In both cases, there will be an increase in the local segment density of the polymer chains in the interaction region. This is schematically illustrated in Figure 8.1. The real situation is perhaps in between the above two cases, that is, the polymer chains may undergo some interpenetration and some compression.

Provided the dangling chains (the A chains in A–B, A–B–A block or BA_n graft copolymers) are in a good solvent, this local increase in segment density in the interaction zone will result in strong repulsion as a result of two main effects [1]. (i) Increase in the osmotic pressure in the overlap region as a result of the unfavorable mixing of the polymer chains, when these are in good solvent conditions. This is referred to as osmotic repulsion or mixing interaction and it is described by a free energy of interaction G_{mix}. (ii) Reduction of the configurational entropy of the chains in the interaction zone. This entropy reduction results from the decrease in the volume available for the chains when these are either overlapped or compressed. This is referred to as volume restriction interaction, entropic or elastic interaction and is described by a free energy of interaction G_{el}.

Combination of G_{mix} and G_{el} is usually referred to as the steric interaction free energy, G_s, that is,

$$G_s = G_{mix} + G_{el} \qquad (8.1)$$

The sign of G_{mix} depends on the solvency of the medium for the chains. If in a good solvent, that is, the Flory–Huggins interaction parameter χ is less than 0.5, then G_{mix} is positive and the mixing interaction leads to repulsion (see below). In contrast if $\chi > 0.5$ (i.e., the chains are in a poor solvent condition), G_{mix} is negative and the mixing interaction becomes attractive. G_{el} is always positive and hence in some cases one can produce stable dispersions in a relatively poor solvent (enhanced steric stabilization).

8.2.1
Mixing Interaction G_{mix}

This results from the unfavorable mixing of the polymer chains, when these are in good solvent conditions. This is schematically shown in Figure 8.2. Consider

8.2 Interaction between Particles Containing Adsorbed Polymeric Surfactant Layers

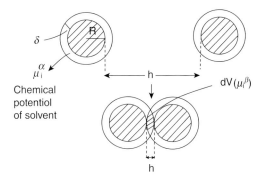

Figure 8.2 Schematic representation of polymer layer overlap.

two spherical particles with the same radius and each containing an adsorbed polymer layer with thickness δ. Before overlap, one can define in each polymer layer a chemical potential for the solvent μ_i^α and a volume fraction for the polymer in the layer ϕ_2. In the overlap region (volume element dV), the chemical potential of the solvent is reduced to μ_i^β. This results from the increase in polymer segment concentration in this overlap region [2–4].

In the overlap region, the chemical potential of the polymer chains is now higher than in the rest of the layer (with no overlap). This amounts to an increase in the osmotic pressure in the overlap region; as a result, solvent will diffuse from the bulk to the overlap region, thus separating the particles, and hence a strong repulsive energy arises from this effect. The above repulsive energy can be calculated by considering the free energy of mixing of two polymer solutions, as for example treated by Flory and Krigbaum [5]. The free energy of mixing is given by two terms: (i) an entropy term that depends on the volume fraction of polymer and solvent; (ii) an energy term that is determined by the Flory–Huggins interaction parameter:

$$\delta(G_{mix}) = kT(n_1 \ln \varphi_1 + n_2 \ln \varphi_2 + \chi n_1 \varphi_2) \tag{8.2}$$

where n_1 and n_2 are the number of moles of solvent and polymer with volume fractions φ_1 and φ_2, k is the Boltzmann constant, and T is the absolute temperature.

The total change in free energy of mixing for the whole interaction zone, V, is obtained by summing over all the elements in V:

$$G_{mix} = \frac{2kTV_2^2}{V_1} v_2 \left(\frac{1}{2} - \chi\right) R_{mix}(h) \tag{8.3}$$

where V_1 and V_2 are the molar volumes of solvent and polymer, respectively, v_2 is the number of chains per unit area and $R_{mix}(h)$ is geometric function that depends on the form of the segment density distribution of the chain normal to the surface, $\rho(z)$. k is the Boltzmann constant and T is the absolute temperature.

Using the above theory one can derive an expression for the free energy of mixing of two polymer layers (assuming a uniform segment density distribution in each layer) surrounding two spherical particles as a function of the separation distance h between the particles [5].

The expression for G_{mix} is

$$G_{mix} = \left(\frac{2V_2^2}{V_1}\right) v_2 \left(\frac{1}{2} - \chi\right)\left(\delta - \frac{h}{2}\right)^2 \left(3R + 2\delta + \frac{h}{2}\right) \quad (8.4)$$

The sign of G_{mix} depends on the value of the Flory–Huggins interaction parameter χ: if $\chi < 0.5$, G_{mix} is positive and the interaction is repulsive; if $\chi > 0.5$, G_{mix} is negative and the interaction is attractive. The condition $\chi = 0.5$ and $G_{mix} = 0$ is termed the θ-condition. The latter corresponds to the case where the polymer mixing behaves as ideal, that is, mixing of the chains does not lead to increase or decrease of the free.

8.2.2
Elastic Interaction G_{el}

This arises from the loss in configurational entropy of the chains on the approach of a second particle. As a result of this approach, the volume available for the chains becomes restricted, resulting in loss of the number of configurations. This can be illustrated by considering a simple molecule, represented by a rod that rotates freely in a hemisphere across a surface (Figure 8.3). When the two surfaces are separated by an infinite distance ∞, the number of configurations of the rod is $\Omega(\infty)$ which is proportional to the volume of the hemisphere. When a second particle approaches to a distance h such that it cuts the hemisphere (loosing some volume), the volume available to the chains is reduced and the number of configurations become $\Omega(h)$ that is less than $\Omega(\infty)$. For two flat plates, G_{el} is given by the following expression:

$$\frac{G_{el}}{kT} = 2v_2 \ln\left[\frac{\Omega(h)}{\Omega(\infty)}\right] = 2v_2 R_{el}(h) \quad (8.5)$$

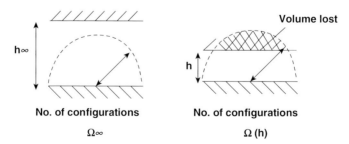

Figure 8.3 Schematic representation of configurational entropy loss on approach of a second particle.

where $R_{el}(h)$ is a geometric function whose form depends on the segment density distribution. It should be stressed that G_{el} is always positive and could play a major role in steric stabilization. It becomes very strong when the separation distance between the particles becomes comparable to the adsorbed layer thickness δ.

8.2.3
Total Energy of Interaction

Combination of G_{mix} and G_{el} with G_A gives the total energy of interaction G_T (assuming there is no contribution from any residual electrostatic interaction), that is,

$$G_T = G_{mix} + G_{el} + G_A \tag{8.6}$$

A schematic of the variation of G_{mix}, G_{el}, G_A, and G_T with surface–surface separation distance h is shown in Figure 8.4. G_{mix} increases very sharply with decrease in h, when $h < 2\delta$. G_{el} increases very sharply with decrease in h, when $h < \delta$. G_T versus h shows a minimum, G_{min}, at separation distances comparable to 2δ. When $h < 2\delta$, G_T shows a rapid increase with decrease in h. The depth of the minimum depends on the Hamaker constant A, the particle radius R, and adsorbed layer thickness δ. G_{min} increases with increase of A and R. At a given A and R, G_{min} increases with decrease in δ (i.e., with decrease of the molecular weight, M_w, of the stabilizer). This is illustrated in Figure 8.5 that shows the energy–distance curves as a function of δ/R. The larger the value of δ/R, the smaller the value of G_{min}. In this case, the system may approach thermodynamic stability as is the case with nano-dispersions.

8.2.4
Criteria for Effective Steric Stabilization

i) The particles should be completely covered by the polymer (the amount of polymer should correspond to the plateau value). Any bare patches may cause flocculation either by van der Waals attraction (between the bare patches) or

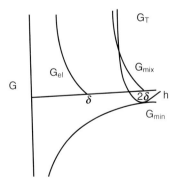

Figure 8.4 Energy–distance curves for sterically stabilized systems.

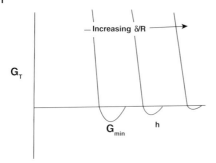

Figure 8.5 Variation of G_{min} with δ/R.

by bridging flocculation (whereby a polymer molecule will become simultaneously adsorbed on two or more particles).

ii) The polymer should be strongly "anchored" to the particle surfaces, to prevent any displacement during particle approach. This is particularly important for concentrated suspensions. For this purpose A–B, A–B–A block and BA_n graft copolymers are the most suitable where the chain B is chosen to be highly insoluble in the medium and has a strong affinity to the surface. Examples of B groups for hydrophobic particles in aqueous media are polystyrene and polymethylmethacrylate.

iii) The stabilizing chain A should be highly soluble in the medium and strongly solvated by its molecules. Examples of A chains in aqueous media are poly(ethylelene oxide) and poly(vinyl alcohol).

iv) δ should be sufficiently large (>10 nm) to prevent weak flocculation.

8.3
Flocculation of Sterically Stabilized Dispersions

Two main types of flocculation may be distinguished.

8.3.1
Weak Flocculation

This occurs when the thickness of the adsorbed layer is small (usually <5 nm), particularly when the particle radius and Hamaker constant are large. The minimum depth required for causing weak flocculation depends on the volume fraction of the suspension. The higher the volume fraction, the lower the minimum depth required for weak flocculation. This can be understood if one considers the free energy of flocculation that consists of two terms, an energy term determined by the depth of the minimum (G_{min}) and an entropy term that is determined by reduction in configurational entropy on aggregation of particles:

$$\Delta G_{flocc} = \Delta H_{flocc} - T\Delta S_{flocc} \tag{8.7}$$

With dilute suspension, the entropy loss on flocculation is larger than with concentrated suspensions. Hence, for flocculation of a dilute suspension, a higher energy minimum is required when compared with the case with concentrated suspensions.

The above flocculation is weak and reversible, that is, on shaking the container redispersion of the suspension occurs. On standing, the dispersed particles aggregate to form a weak "gel." This process (referred to as sol ↔ gel transformation) leads to reversible time dependence of viscosity (thixotropy). On shearing the suspension, the viscosity decreases and when the shear is removed, the viscosity is recovered. This phenomenon is applied in paints. On application of the paint (by a brush or roller), the gel is fluidized, allowing uniform coating of the paint. When shearing is stopped, the paint film recovers its viscosity and this avoids any dripping.

8.3.2
Incipient Flocculation

This occurs when the solvency of the medium is reduced to become worse than θ-solvent (i.e., $\chi > 0.5$). This is illustrated in Figure 8.6 where χ was increased from <0.5 (good solvent) to >0.5 (poor solvent).

When $\chi > 0.5$, G_{mix} becomes negative (attractive) which when combined with the van der Waals attraction at this separation distance gives a deep minimum causing flocculation. In most cases, there is a correlation between the critical flocculation point and the θ condition of the medium. Good correlation is found in many cases between the critical flocculation temperature (CFT) and θ-temperature of the polymer in solution (with block and graft copolymers one should consider the θ-temperature of the stabilizing chains A) [6]. Good correlation is also found

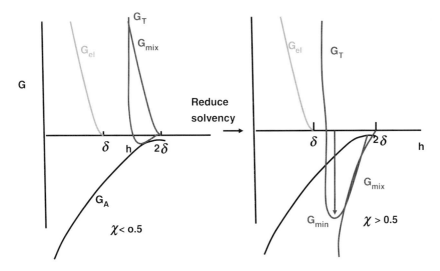

Figure 8.6 Influence of reduction in solvency on the energy–distance curve.

between the critical volume fraction (CFV) of a nonsolvent for the polymer chains and their θ-point under these conditions. However, in some cases such correlation may break down, particularly the case for polymers that adsorb by multipoint attachment. This situation has been described by Napper [6] who referred to it as "enhanced" steric stabilization.

Thus, by measuring the θ-point (CFT or CFV) for the polymer chains (A) in the medium under investigation (that could be obtained from viscosity measurements) one can establish the stability conditions for a dispersion, before its preparation. This procedure helps also in designing effective steric stabilizers such as block and graft copolymers.

8.3.3
Depletion Flocculation

Depletion flocculation is produced by addition of "free" nonadsorbing polymer [7]. In this case, the polymer coils cannot approach the particles to a distance Δ (that is determined by the radius of gyration of free polymer R_G), since the reduction of entropy on close approach of the polymer coils is not compensated by the adsorption energy. The suspension particles will be surrounded by a depletion zone with thickness Δ. Above a critical volume fraction of the free polymer, φ_p^+, the polymer coils are "squeezed out" from between the particles and the depletion zones begin to interact. The interstices between the particles are now free from polymer coils and hence an osmotic pressure is exerted outside the particle surface (the osmotic pressure outside is higher than in between the particles) resulting in weak flocculation [7]. A schematic representation of depletion flocculation is shown in Figure 8.7.

The magnitude of the depletion attraction free energy, G_{dep}, is proportional to the osmotic pressure of the polymer solution, which in turn is determined by φ_p and molecular weight M. The range of depletion attraction is proportional to the thickness of the depletion zone, Δ, which is roughly equal to the radius of gyration, R_G, of the free polymer. A simple expression for G_{dep} is [7],

$$G_{dep} = \frac{2\pi R \Delta^2}{V_1}(\mu_1 - \mu_1^0)\left(1 + \frac{2\Delta}{R}\right) \tag{8.8}$$

where V_1 is the molar volume of the solvent, μ_1 is the chemical potential of the solvent in the presence of free polymer with volume fraction φ_p and μ_1^0 is the chemical potential of the solvent in the absence of free polymer. $(\mu_1 - \mu_1^0)$ is proportional to the osmotic pressure of the polymer solution.

8.4
Bridging Flocculation by Polymers and Polyelectrolytes

Certain long-chain polymers may adsorb in such a way that different segments of the same polymer chain are adsorbed on different particles, thus binding or "bridg-

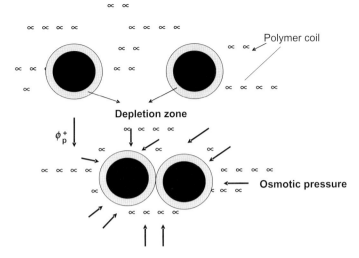

Figure 8.7 Schematic representation of depletion flocculation.

ing" the particles together, despite the electrical repulsion [8]. With polyelectrolytes of opposite charge to the particles, another possibility exists; the particle charge may be partly or completely neutralized by the adsorbed polyelectrolyte, thus reducing or eliminating the electrical repulsion and destabilizing the particles.

Effective flocculants are usually linear polymers, often of high molecular weight, which may be nonionic, anionic, or cationic in character. Ionic polymers should strictly be referred to as polyelectrolytes. The most important properties are molecular weight and charge density. There are several polymeric flocculants that are based on natural products, for example, starch and alginates, but the most commonly used flocculants are synthetic polymers and polyelectrolytes, for example, polacrylamide and copolymers of acrylamide and a suitable cationic monomer such as dimethylaminoethyl acrylate or methacrylate. Other synthetic polymeric flocculants are poly(vinyl alcohol), poly(ethylene oxide) (nonionic), sodium polystyrene sulfonate (anionic), and polyethyleneimine (cationic).

As mentioned above, bridging flocculation occurs because segments of a polymer chain adsorb simultaneously on different particles thus linking them together. Adsorption is an essential step and this requires favorable interaction between the polymer segments and the particles. Several types of interactions are responsible for adsorption that is irreversible in nature. (i) Electrostatic interaction when a polyelectrolyte adsorbs on a surface bearing oppositely charged ionic groups, for example, adsorption of a cationic polyelectrolyte on a negative oxide surface such as silica. (ii) Hydrophobic bonding that is responsible for adsorption of nonpolar segments on a hydrophobic surface, for example, partially hydrolyzed poly(vinyl acetate) (PVA) on a hydrophobic surface such as polystyrene. (iii) Hydrogen bonding as for example interaction of the amide group of polyacrylamide with hydroxyl groups on an oxide surface. (iv) Ion binding as is the case of

Figure 8.8 Schematic illustration of bridging flocculation (a) and restabilization (b) by the adsorbed polymer.

adsorption of anionic polyacrylamide on a negatively charged surface in the presence of Ca^{2+}.

Effective bridging flocculation requires the adsorbed polymer extends far enough from the particle surface to attach to other particles and that there is sufficient free surface available for adsorption of these segments of extended chains. When excess polymer is adsorbed, the particles can be restabilized, either because of surface saturation or by steric stabilization as discussed earlier. This is one explanation of the fact that an "optimum dosage" of flocculant is often found; at low concentration there is insufficient polymer to provide adequate links and with larger amounts restabilization may occur. A schematic of bridging flocculation and restabilization by adsorbed polymer is given in Figure 8.8.

If the fraction of the particle surface covered by polymer is θ, then the fraction of uncovered surface is $(1 - \theta)$ and the successful bridging encounters between the particles should be proportional to $\theta(1 - \theta)$, which has its maximum when $\theta = 0.5$. This is the well-known condition of "half-surface-coverage" that has been suggested as giving the optimum flocculation.

An important condition for bridging flocculation with charged particles is the role of electrolyte concentration. The latter determines the extension ("thickness") of the double layer that can reach values as high as 100 nm (in 10^{-5} mol dm^{-5} 1:1 electrolyte such as NaCl). For bridging flocculation to occur, the adsorbed polymer must extend far enough from the surface to a distance over which electrostatic repulsion occurs (>100 nm in the above example). This means that at low electrolyte concentrations quite high molecular weight polymers are needed for bridging to occur. As the ionic strength is increased, the range of electrical repulsion is reduced and lower molecular weight polymers should be effective.

In many practical applications, it has been found that the most effective flocculants are polyelectrolytes with a charge opposite to that of the particles. In aqueous media most particles are negatively charged, and cationic polyelectrolytes such as polyethyleneimine are often necessary. With oppositely charged polyelectrolytes, it is likely that adsorption occurs to give a rather flat configuration of the adsorbed chain, due to the strong electrostatic attraction between the positive ionic groups on the polymer and the negatively charged sites on the particle surface. This would probably reduce the probability of bridging contacts with other particles, especially with fairly low molecular weight polyelectrolytes with high charge density. However, the adsorption of a cationic polyelectrolyte on a negatively charged particle will reduce the surface charge of the latter, and this charge neutralization could be an important factor in destabilizing the particles. Another mechanism for destabilization has been suggested by Gregory [8] who proposed an "electrostatic-patch" model. This applied to cases where the particles have a fairly low density of immobile charges and the polyelectrolyte has a fairly high charge density. Under these conditions, it is not physically possible for each surface site to be neutralized by a charged segment on the polymer chain, even though the particle may have sufficient adsorbed polyelectrolyte to achieve overall neutrality. There are then "patches" of excess positive charge, corresponding to the adsorbed polyelectrolyte chains (probably in a rather flat configuration), surrounded by areas of negative charge, representing the original particle surface. Particles that have this "patchy" or "mosaic" type of surface charge distribution may interact in such a way that the positive and negative "patches" come into contact, giving quite strong attraction (although not as strong as in the case of bridging flocculation). A schematic of this type of interaction is given in Figure 8.9. The electrostatic-patch concept (that can be regarded as another form of "bridging") can explain a number of features of flocculation of negatively charged particles with positive polyelectrolytes. These include the rather small effect of increasing the molecular weight and the effect of ionic strength on the breadth of the flocculation dosage range and the rate of flocculation at optimum dosage.

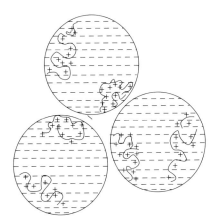

Figure 8.9 "Electrostatic-patch" model for the interaction of negatively charged particles with adsorbed cationic polyelectrolytes.

8.5
Examples for Suspension Stabilization Using Polymeric Surfactants

There are generally two procedures for preparation of solid–liquid dispersions.

i) **Condensation methods**: build-up of particles from molecular units, that is, nucleation and growth. A special procedure is the preparation of latexes by emulsion or dispersion polymerization.

ii) **Dispersion methods**: in this case, one starts with preformed large particles or crystals that are dispersed in the liquid by using a surfactant (wetting agent) with subsequent breaking up of the large particles by milling (comminution) to achieve the desirable particle size distribution. A dispersing agent (usually a polymeric surfactant) is used for the dispersion process and subsequent stabilization of the resulting suspension.

There are generally two procedures for preparation of latexes.

i) **Emulsion polymerization**: the monomers that are essentially insoluble in the aqueous medium are emulsified using a surfactant and an initiator is added while heating the system to produce the polymer particles that are stabilized electrostatically (when using ionic surfactants) or sterically (when using nonionic surfactants.

ii) **Dispersion polymerization**: the monomers are dissolved in a solvent in which the resulting polymer particles are insoluble. A protective colloid (normally a block or graft copolymer) is added to prevent flocculation of the resulting polymers particles that are produced on addition of an initiator. This method is usually applied for the production of nonaqueous latex dispersions and is sometimes referred to as nonaqueous dispersion polymerization (NAD).

Recently, the graft copolymer of hydrophobically modified inulin (INUTEC® SP1) has been used in emulsion polymerization of styrene, methyl methacrylate, butyl acrylate, and several other monomers [9]. All lattices were prepared by emulsion polymerization using potassium persulphate as initiator. The z-average particle size was determine by photon correlation spectroscopy (PCS) and electron micrographs were also taken.

Emulsion polymerization of styrene or methyl methacrylate showed an optimum ratio of (INUTEC)/Monomer of 0.0033 for PS and 0.001 for PMMA particles. The (initiator) to (monomer) ratio was kept constant at 0.00125. The monomer conversion was higher than 85% in all cases. Latex dispersions of PS reaching 50% and of PMMA reaching 40% could be obtained using such low concentration of INUTEC®SP1. Figure 8.10 shows the variation of particle diameter with monomer concentration.

The stability of the latexes was determined by determining the critical coagulation concentration (CCC) using $CaCl_2$. The CCC was low (0.0175–0.05 mol dm^{-3}) but this was higher than that for the latex prepared without surfactant. Postaddition of INUTEC®SP1 resulted in a large increase in the CCC as is illustrated in Figure 8.11 that shows log W–log C curves (where W is the stability ratio) at various additions of INUTEC®SP1.

8.5 Examples for Suspension Stabilization Using Polymeric Surfactants | 143

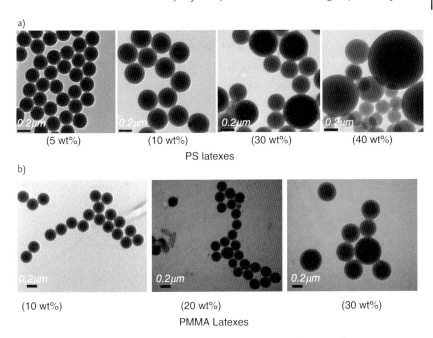

Figure 8.10 Electron micrographs of the latexes. (a) PS latexes. (b) PMMA latexes.

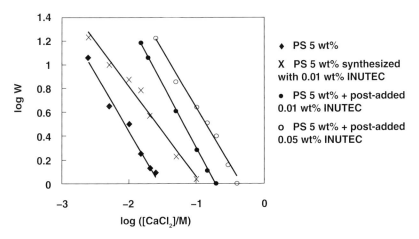

Figure 8.11 Influence of post addition of INUTEC®SP1 on the latex stability.

The high stability of the latex when using INUTEC®SP1 is due to the strong adsorption of the polymeric surfactant on the latex particles and formation of strongly hydrated loops and tails of polyfructose that provide effective steric stabilization. Evidence for the strong repulsion produced when using INUTEC®SP1 was obtained from atomic force microscopy investigations [10] whereby the force between hydrophobic glass spheres and hydrophobic glass plate, both containing an adsorbed layer of INUTEC®SP1, was measured as a function of distance of separation both in water and in the presence of various Na_2SO_4 concentrations. The AFM used was capable of measuring picoNewton surface forces at nanometer length scales. The interaction between glass spheres that were attached to the AFM cantilever and glass plates (both hydrophobized using dichlorodimethylsilane) and containing adsorbed layers of hydrophobically modified inulin (INUTEC SP1) was measured as a function of INUTEC SP1 concentration in water and at various Na_2SO_4 concentrations.

Measurements were initially carried out as a function of time (2, 5, and 24 h) at 2×10^{-4} mol dm^{-3} INUTEC SP1. The force–separation distance curves showed that after 2 and 5 h equilibration time, the force–separation distance curve showed some residual attraction on withdrawal. By increasing the equilibration time (24 h) this residual attraction on withdrawal disappeared and the approach and withdrawal curves were very close indicating full coverage of the surfaces with polymer within this time. All subsequent measurements were carried out after 24 h to ensure complete adsorption. Measurements were carried out at five different concentrations of INUTEC SP1: 6.6×10^{-6}, 1×10^{-5}, 6×10^{-5}, 1.6×10^{-4}, and 2×10^{-4} mol dm^{-3}. At concentrations <1.6×10^{-4} mol dm^{-3} the withdrawal curve showed some residual attraction as is illustrated in Figure 8.12 for 6×10^{-5} mol dm^{-3}. At concentrations >1.6×10^{-4} mol dm^{-3} the approach and withdrawal curves were very close to each other as is illustrated in Figure 8.13 for 2×10^{-4} mol dm^{-3}. These results indicate full coverage of the surfaces by the polymer when the INUTEC SP1 concentration becomes equal or higher than 1.6×10^{-4} mol dm^{-3}. The results at full coverage give an adsorbed layer thickness of the order of 9 nm that indicate strong hydration of the loops and tails of inulin [10].

Several investigations of the stability of emulsions and suspensions stabilized using INUTEC SP1 showed the absence of flocculation in the presence of high electrolyte concentrations (up to 4 mol dm^{-3} NaCl and 1.5 mol dm^{-3} MgSO$_4$). This

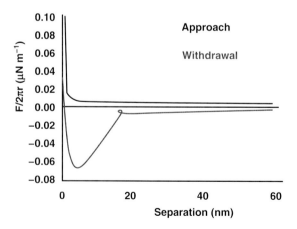

Figure 8.12 Force–distance curves at 6×10^{-5} mol dm^{-3} INUTEC SP1.

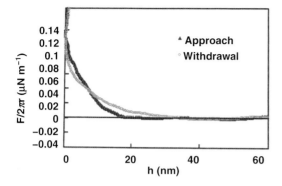

Figure 8.13 Force–distance curves at $2 \times 10^{-4}\,\text{mol}\,\text{dm}^{-3}$ INUTEC SP1.

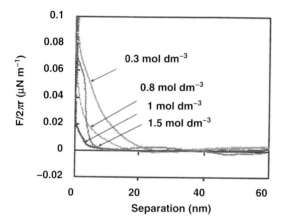

Figure 8.14 Force–distance curves for hydrophobized glass surfaces containing adsorbed INUTEC®SP1 at various Na_2SO_4 concentrations.

high stability in the presence of high electrolyte concentrations is attributed to the strong hydration of inulin (polyfructose) loops and tails. This strong hydration was confirmed by measuring the cloud point of inulin in the presence of such high electrolyte concentrations (the cloud point exceeded 100 °C up to $4\,\text{mol}\,\text{dm}^{-3}$ NaCl and $1.0\,\text{mol}\,\text{dm}^{-3}$ $MgSO_4$). Evidence of such strong repulsion was obtained from the force–distance curves in the presence of different concentrations of Na_2SO_4 as shown in Figure 8.14.

The force–distance curves clearly show that the interaction remains repulsive up to the highest Na_2SO_4 concentration ($1.5\,\text{mol}\,\text{dm}^{-3}$) studied. The adsorbed layer thickness decreases from approximately 9 nm at $0.3\,\text{mol}\,\text{dm}^{-3}$ to about 3 nm at $1.5\,\text{mol}\,\text{dm}^{-3}$ Na_2SO_4. This reduction in hydrodynamic thickness in the presence of high electrolyte concentrations is probably due to the change in the conformation of polyfructose loops and tails. It is highly unlikely that dehydration of the chains occurs since cloud point measurements have shown absence of any cloud point up to 100 °C. Even at such low adsorbed layer thickness, strong repulsive interaction is observed indicating a high elastic repulsive term.

Dispersion polymerization: this method is usually applied for the preparation of nonaqueous latex dispersions and hence it is referred to as NAD. The method has also been adapted to prepare aqueous latex dispersions by using an alcohol–water mixture. In the NAD process, the monomer, normally an acrylic, is dissolved in a nonaqueous solvent, normally an aliphatic hydrocarbon and an oil soluble initiator and a stabilizer (to protect the resulting particles from flocculation) is added to the reaction mixture. The most successful stabilizers used in NAD are block and graft copolymers. Preformed graft stabilizers based on poly(12-hydroxy stearic acid) (PHS) are simple to prepare and effective in NAD polymerization. Commercial 12-hydroxystearic acid contains 8–15% palmitic and stearic acids that limits the molecular weight during polymerization to an average of 1500–2000. This oligomer may be converted to a "macromonomer" by reacting the carboxylic group with glycidyl methacrylate. The macromonomer is then copolymerized with an equal weight of methyl methacrylate (MMA) or similar monomer to give a "comb" graft copolymer with an average molecular weight of 10 000–20 000. The graft copolymer contains on average 5–10 PHS chains pendent from a polymeric anchor backbone of PMMA. This graft copolymer can stabilize latex particles of various monomers. The major limitation of the monomer composition is that the polymer produced should be insoluble in the medium used.

NAD polymerization is carried in two steps. (i) Seed stage: the diluent, portion of the monomer, portion of dispersant and initiator (azo or peroxy type) are heated to form an initial low-concentration fine dispersion. (ii) Growth stage: the remaining monomer together with more dispersant and initiator are then fed over the course of several hours to complete the growth of the particles. A small amount of transfer agent is usually added to control the molecular weight. Excellent control of the particle size is achieved by proper choice of the designed dispersant and correct distribution of dispersant between the seed and growth stages. NAD acrylic polymers are applied in automotive thermosetting polymers and hydroxy monomers may be included in the monomer blend used.

8.6
Polymeric Surfactants for Stabilization of Preformed Latex Dispersions

For this purpose polystyrene (PS) latexes were prepared using the surfactant-free emulsion polymerization. Two latexes with z-average diameter of 427 and 867 (as measured using photon correlation spectroscopy (PCS)) that are reasonably monodisperse were prepared [11]. Two polymeric surfactants, namely Hypermer CG-6 and Atlox 4913 were used. Both are graft ("comb") type consisting of polymethylmethacrylate/polymethacrylic acid (PMMA/PMA) backbone with methoxy-capped polyethylene oxide (PEO) side chains ($M = 750$ Da). Hypermer

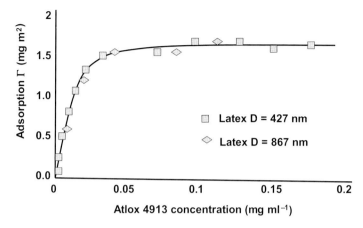

Figure 8.15 Adsorption isotherms of Atlox 4913 on the two latexes at 25 °C.

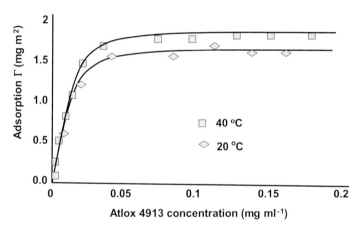

Figure 8.16 Effect of temperature on adsorption of Atlox 4913 on PS.

CG-6 is the same graft copolymer as Atlox 4913 but it contains higher proportion of methacrylic acid in the backbone. The average molecular weight of the polymer is ~ 5000 Da. Figure 8.15 shows a typical adsorption isotherm of Atlox 4913 on the two latexes. Similar results were obtained for Hypermer CG-6 but the plateau adsorption was lower (1.2 mg m^{-2} compared with 1.5 mg m^{-2} for Atlox 4913). It is likely that the backbone of Hypermer CG-6 that contains more PMA is more polar and hence less strongly adsorbed. The amount of adsorption was independent of the particle size.

The influence of temperature on adsorption is shown in Figure 8.16. The amount of adsorption increases with increase of temperature. This is due to the poorer solvency of the medium for the PEO chains. The PEO chains become less

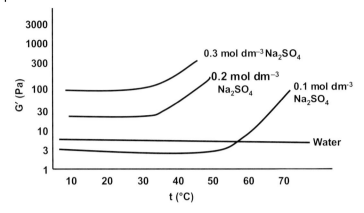

Figure 8.17 Variation of G' with temperature in water and at various Na_2SO_4 concentrations.

hydrated at higher temperature and the reduction of solubility of the polymer enhances adsorption.

The latex stability was assessed using viscoelastic measurements (see Chapter 12 on rheology) where the elastic modulus G' is measured as a function of electrolyte concentration and/or temperature to assess the latex stability. As an illustration Figure 8.17 shows the variation of G' with temperature for latex stabilized with Atlox 4913 in the absence of any added electrolyte and in the presence of 0.1, 0.2, and 0.3 mol dm^{-3} Na_2SO_4. In the absence of electrolyte G' showed no change with temperature up to 65 °C.

In the presence of 0.1 mol dm^{-3} Na_2SO_4, G' remained constant up to 40 °C above which G' increased with further increase of temperature. This temperature is denoted as the CFT. The CFT decreases with increase in electrolyte concentration reaching ~30 °C in 0.2 and 0.3 mol dm^{-3} Na_2SO_4. This reduction in CFT with increase in electrolyte concentration is due to the reduction in solvency of the PEO chains with increase in electrolyte concentrations. The latex stabilized with Hypermer CG-6 gave relatively higher CFT values when compared with that stabilized using Atlox 4913.

References

1 Tadros, T.F. (1981) Polymer adsorption and dispersion stability, in *The Effect of Polymers on Dispersion Properties*, (ed. T.F. Tadros), Academic Press, London.
2 Fischer, E.W. (1958) *Kolloid Z.*, **160**, 120.
3 Mackor, E.L., and van der Waals, J.H. (1951) *J. Colloid Sci.*, **7**, 535.
4 Hesselink, F.T., Vrij, A., and Overbeek, J.T.G. (1971) *J. Phys. Chem.*, **75**, 2094.
5 Flory, P.J., and Krigbaum, W.R. (1950) *J. Chem. Phys.*, **18**, 1086.
6 Napper, D.H. (1981) *Polymeric Stabilisation of Colloidal Dispersions*, Academic Press, London.
7 Asakura, A., and Oosawa, F. (1954) *J. Chem. Phys.*, **22**, 1235; Asakura, A., and Oosawa, F. (1958) *J. Polym. Sci.*, **93**, 183.

8 Gregory, J. (1987) *Solid/Liquid Dispersions*, (ed. T.F. Tadros), Academic Press, London.
9 Nestor, J., Esquena, J., Solans, C., Levecke, B., Booten, K., and Tadros, T.F. (2005) *Langmuir*, **21**, 4837.
10 Nestor, J., Esquena, J., Solans, C., Luckham, P.F., Levecke, B., and Tadros, T.F. (2007) *J. Colloid Interface Sci.*, **311**, 430.
11 Liang, W., Bognolo, G., and Tadros, T.F. (1995) *Langmuir*, **11**, 2899.

9
Properties of Concentrated Suspensions

One of the main features of concentrated suspensions is the formation of three-dimensional structure units, which determine their properties and in particular their rheology. The formation of these units is determined by the interparticle interactions, which need to be clearly defined and quantified. These interparticle interactions have been described in detail in Chapters 4 and 8 and will be summarized in this chapter. The second part will give a definition of the particle number concentration above which a suspension may be considered concentrated. The various states of the suspension that are reached are then described and analyzed in terms of the interparticle interactions and the effect of gravity.

9.1
Interparticle Interactions and Their Combination

For the control of the properties of concentrated suspensions one has to control the interparticle interactions [1–3]. Four different types of interactions can be distinguished as illustrated in Figure 9.1 and these are summarized below:

9.1.1
Hard-Sphere Interaction

These are systems where both repulsion and attraction have been screened (Figure 9.1a). A good example is aqueous polystyrene latex with a radius $R > 100\,nm$ whereby the double layers are compressed by addition of 1:1 electrolyte such as NaCl at a concentration of $10^{-2}\,mol\,dm^{-3}$ whereby the double layer thickness is 3.3 nm. Alternatively the double layer is screened by replacing water with a less polar medium such as benzyl alcohol. The particles are considered to behave as "hard-spheres" with a radius R_{HS} that is slightly larger than the core radius R. When the particles reach a distance a center-to-center distance that is smaller than $2R_{HS}$ the interaction increases very sharply approaching ∞. One can define a maximum hard-sphere volume fraction above which the flow behavior suddenly changes from "fluid-like" to "solid-like" (viscous to elastic response).

Dispersion of Powders in Liquids and Stabilization of Suspensions, First Edition. Tharwat F. Tadros.
© 2012 Wiley-VCH Verlag GmbH & Co. KGaA. Published 2012 by Wiley-VCH Verlag GmbH & Co. KGaA.

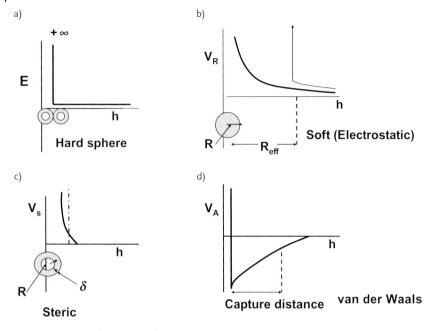

Figure 9.1 Types of interaction forces.

Table 9.1 Values of $(1/\kappa)$ for 1:1 electrolyte at 25 °C.

C (mol dm^{-3})	10^{-5}	10^{-4}	10^{-3}	10^{-2}	10^{-1}
$(1/\kappa)$ (nm)	100	33	10	3.3	1

9.1.2
"Soft" or Electrostatic Interaction: Figure 9.1b

The particles in this case have a surface charge (either by ionization of surface groups as in the case of oxides) or in the presence of adsorbed ionic surfactants. The surface charge σ_0 is compensated by unequal distribution of counterions (opposite in charge to the surface) and co-ions (same sign as the surface) that extend to some distance from the surface [4, 5]. The double layer extension depends on electrolyte concentration and valency of the counterions:

$$\left(\frac{1}{\kappa}\right) = \left(\frac{\varepsilon_r \varepsilon_0 k T}{2 n_0 Z_i^2 e^2}\right)^{1/2} \tag{9.1}$$

ε_r is the permittivity (dielectric constant); 78.6 for water at 25 °C. ε_0 is the permittivity of free space. k is the Boltzmann constant and T is the absolute temperature. n_0 is the number of ions per unit volume of each type present in bulk solution and Z_i is the valency of the ions and e is the electronic charge.

Values of $(1/\kappa)$ for 1:1 electrolyte (e.g., KCl) are given in Table 9.1.

The double layer extension increases with decrease in electrolyte concentration.

When charged colloidal particles in a dispersion approach each other such that the double layers begin to overlap (particle separation becomes less than twice the double layer extension), repulsion occurs. The individual double layers can no longer develop unrestrictedly, since the limited space does not allow complete potential decay [4–6].

For two spherical particles of radius R and surface potential ψ_0 and condition $\kappa R < 3$, the expression for the electrical double layer repulsive interaction is given by [4]

$$G_{el} = \frac{4\pi\varepsilon_r\varepsilon_0 R^2 \psi_0^2 \exp-(\kappa h)}{2R+h} \qquad (9.2)$$

where h is the closest distance of separation between the surfaces.

The above expression shows the exponential decay of G_{el} with h. The higher the value of κ (i.e., the higher the electrolyte concentration), the steeper the decay. This means that at any given distance h, the double layer repulsion decreases with increase of electrolyte concentration.

The importance of the double layer extension can be illustrated as follows. Let us consider a very small particle with a radius of 10 nm in 10^{-5} mol dm^{-3} NaCl. The core radius $R = 10$ nm but the effective radius R_{eff} (the core radius plus the double layer thickness) is now 110 nm. The core volume is $(4/3)\pi R^3 = (4/3)\pi(10)^3$ nm^3 but the effective volume is now $= (4/3)\pi(110)^3$ which is ~1000 times higher than the core volume. The same applies to the volume fraction:

$$\phi_{eff} = \phi\left(\frac{110}{10}\right)^3 \approx 1000\phi \qquad (9.3)$$

For a dispersion with equal size particles, the maximum possible volume fraction (for hexagonal packing) ϕ_p is 0.74. For the above case of 10 nm particles in 10^{-5} mol dm^{-3} NaCl the maximum core volume fraction is ~7.4×10^{-4} and at above this volume fraction strong repulsion occurs and the system may now be considered concentrated.

With increase in electrolyte concentration, the double layer thickness decreases and ϕ_{eff} decreases and the core volume fraction has to be increased to produce a "concentrated" system.

9.1.3
Steric Interaction: Figure 9.1c

This occurs when the particles contain adsorbed nonionic surfactant or polymer layers of the A–B, A–B–A block or BA$_n$ graft types, where B is the "anchor" chain that has a high affinity to the surface (strongly adsorbed) and A is the "stabilizing" chain that is highly soluble in the medium and strongly solvated by its molecules. One can define an adsorbed layer thickness δ_h for the surfactant or polymer and hence an effective radius $R_{eff} = R + \delta_h$.

When two particles each with a radius R and containing an adsorbed polymer layer with a hydrodynamic thickness δ_h, approach each other to a surface–surface separation distance h that is smaller than $2\delta_h$, the polymer layers interact with each other resulting in two main situations [7]: (i) The polymer chains may overlap with each other. (ii) The polymer layer may undergo some compression. In both cases, there will be an increase in the local segment density of the polymer chains in the interaction region. The real situation is perhaps in between the above two cases, that is, the polymer chains may undergo some interpenetration and some compression.

Provided the dangling chains (the A chains in A–B, A–B–A block or BA$_n$ graft copolymers) are in a good solvent, this local increase in segment density in the interaction zone will result in strong repulsion as a result of two main effects. (i) Increase in the osmotic pressure in the overlap region as a result of the unfavorable mixing of the polymer chains, when these are in good solvent conditions. This is referred to as osmotic repulsion or mixing interaction and it is described by a free energy of interaction G_{mix}. (ii) Reduction of the configurational entropy of the chains in the interaction zone. This entropy reduction results from the decrease in the volume available for the chains when these are either overlapped or compressed. This is referred to as volume restriction interaction, entropic, or elastic interaction and it is described by a free energy of interaction G_{el}.

Combination of G_{mix} and G_{el} is usually referred to as the steric interaction free energy, G_s, that is,

$$G_s = G_{mix} + G_{el} \tag{9.4}$$

The sign of G_{mix} depends on the solvency of the medium for the chains. If in a good solvent, that is, the Flory–Huggins interaction parameter χ is less than 0.5, then G_{mix} is positive and the mixing interaction leads to repulsion (see below). In contrast if $\chi > 0.5$ (i.e., the chains are in a poor solvent condition), G_{mix} is negative and the mixing interaction becomes attractive. G_{el} is always positive and hence in some cases one can produce stable dispersions in a relatively poor solvent (enhanced steric stabilization).

As mentioned earlier, the mixing interaction results from the unfavorable mixing of the polymer chains, when these are in good solvent conditions. Consider two spherical particles with the same radius and each containing an adsorbed polymer layer with thickness δ. Before overlap, one can define in each polymer layer a chemical potential for the solvent μ_i^α and a volume fraction for the polymer in the layer ϕ_2. In the overlap region (volume element dV), the chemical potential of the solvent is reduced to μ_i^β. This results from the increase in polymer segment concentration in this overlap region.

In the overlap region, the chemical potential of the polymer chains is now higher than in the remaining layer (with no overlap). This amounts to an increase in the osmotic pressure in the overlap region; as a result, solvent will diffuse from the bulk to the overlap region, thus separating the particles and hence a strong repulsive energy arises from this effect. The above repulsive energy can be calculated by considering the free energy of mixing of two polymer solutions, as for example treated by Flory and Krigbaum [8]. The free energy of mixing is given by two terms. (i) An entropy term that depends on the volume fraction of polymer and solvent. (ii) An energy term that is determined by the Flory–Huggin interaction parameter χ.

Using the above theory, one can derive an expression for the free energy of mixing of two polymer layers (assuming a uniform segment density distribution in each layer) surrounding two spherical particles as a function of the separation distance h between the particles [7]. The expression for G_{mix} is

$$\frac{G_{mix}}{kT} = \left(\frac{2V_2^2}{V_1}\right)v_2^2\left(\frac{1}{2}-\chi\right)\left(\delta-\frac{h}{2}\right)^2\left(3R+2\delta+\frac{h}{2}\right) \quad (9.5)$$

where k is the Boltzmann constant, T is the absolute temperature, V_2 is the molar volume of polymer, V_1 is the molar volume of solvent and v_2 is the number of polymer chains per unit area.

The sign of G_{mix} depends on the value of the Flory–Huggins interaction parameter χ: if $\chi < 0.5$, G_{mix} is positive and the interaction is repulsive; if $\chi > 0.5$, G_{mix} is negative and the interaction is attractive; if $\chi = 0.5$, $G_{mix} = 0$ and this defines the θ condition.

The elastic interaction arises from the loss in configurational entropy of the chains on the approach of a second particle. As a result of this approach, the volume available for the chains becomes restricted, resulting in loss of the number of configurations. This can be illustrated by considering a simple molecule, represented by a rod that rotates freely in a hemisphere across the surface. When the two surfaces are separated by an infinite distance ∞, the number of configurations of the rod is $\Omega(\infty)$ that is proportional to the volume of the hemisphere. When a second particle approaches to a distance h such that it cuts the hemisphere (loosing some volume), the volume available to the chains is reduced and the number of configurations become $\Omega(h)$ that is less than $\Omega(\infty)$. For two flat plates, G_{el} is given by the following expression [7]:

$$\frac{G_{el}}{kT} = 2v_2 \ln\left[\frac{\Omega(h)}{\Omega(\infty)}\right] = 2v_2 R_{el}(h) \quad (9.6)$$

where $R_{el}(h)$ is a geometric function whose form depends on the segment density distribution. It should be stressed that G_{el} is always positive and could play a major role in steric stabilization. It becomes very strong when the separation distance between the particles becomes comparable to the adsorbed layer thickness δ.

One can also define an effective volume fraction ϕ_{eff} that is determined by the ratio of the adsorbed layer thickness δ to the core radius:

$$\phi_{eff} = \phi\left[1+\left(\frac{\delta}{R}\right)\right]^3 \quad (9.7)$$

If (δ/R) is small (say for a particle with radius 1000 nm and δ of 10 nm), $\phi_{eff} \sim \phi$ and the dispersion behaves as near "hard-sphere." In this case, one can reach high ϕ values before the system becomes "concentrated." If (δ/R) is appreciable, say > 0.2 (e.g., for particles with a radius of 100 nm and δ of 20 nm) $\phi_{eff} > \phi$ and the system shows strong interaction at relatively low ϕ values. For example if

$\delta = 0.5\phi$, $\phi_{eff} \sim 3.4\phi$ and the system may be considered concentrated at a core volume fraction of ~0.2.

9.1.4
van der Waals Attraction: Figure 9.1d

As is well known atoms or molecules always attract each other at short distances of separation. The attractive forces are of three different types: Dipole–dipole interaction (Keesom), dipole–induced-dipole interaction (Debye) and London dispersion force. The London dispersion force is the most important, since it occurs for polar and nonpolar molecules. It arises from fluctuations in the electron density distribution.

At small distances of separation r in vacuum, the attractive energy between two atoms or molecules is given by

$$G_{aa} = -\frac{\beta_{11}}{r^6} \tag{9.8}$$

β_{11} is the London dispersion constant.

For colloidal particles that are made of atom or molecular assemblies, the attractive energies may be added and this results in the following expression for two spheres (at small h) [9]:

$$G_A = -\frac{A_{11(2)} R}{12 h} \tag{9.9}$$

where $A_{11(2)}$ is the effective Hamaker constant of two identical particles with Hamaker constant A_{11} in a medium with Hamaker constant A_{22}. When the particles are dispersed in a liquid medium, the van der Waals attraction has to be modified to take into account the medium effect. When two particles are brought from infinite distance to h in a medium, an equivalent amount of medium has to be transported the other way round. Hamaker forces in a medium are excess forces.

The effective Hamaker constant for two identical particles 1 and 1 in a medium 2 is given by

$$A_{11(2)} = A_{11} + A_{22} - 2A_{12} = \left(A_{11}^{1/2} - A_{22}^{1/2}\right)^2 \tag{9.10}$$

Equation (9.8) shows that two particles of the same material attract each other unless their Hamaker constant exactly matches each other. The Hamaker constant of any material is given by

$$A = \pi^2 q^2 \beta_{ii} \tag{9.11}$$

where q is the number of atoms or molecules per unit volume.

In most cases, the Hamaker constant of the particles is higher than that of the medium.

As shown in Figure 9.1d, V_A increases very sharply with h at small h values. A capture distance can be defined at which all the particles become strongly attracted to each other (coagulation).

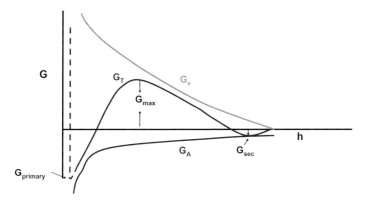

Figure 9.2 Schematic representation of the variation of G_T with h according to the DLVO theory.

9.1.5
Combination of Interaction Forces

Combination of G_{el} and G_A results in the well-known theory of stability of colloids (DLVO Theory) [10, 11]:

$$G_T = G_{el} + G_A \tag{9.12}$$

A plot of G_T versus h is shown in Figure 9.2, which represents the case at low electrolyte concentrations, that is, strong electrostatic repulsion between the particles. G_{el} decays exponentially with h, that is, $G_{el} \to 0$ as h becomes large. G_A is ∞ $1/h$, that is, G_A does not decay to 0 at large h.

At long distances of separation, $G_A > G_{el}$ resulting in a shallow minimum (secondary minimum). At very short distances, $G_A \gg G_{el}$ resulting in a deep primary minimum.

At intermediate distances, $G_{el} > G_A$ resulting in energy maximum, G_{max}, whose height depends on ψ_0 (or ψ_d) and the electrolyte concentration and valency.

At low electrolyte concentrations ($<10^{-2}\,\text{mol dm}^{-3}$ for a 1:1 electrolyte), G_{max} is high ($>25\,kT$) and this prevents particle aggregation into the primary minimum. The higher the electrolyte concentration (and the higher the valency of the ions), the lower the energy maximum.

Under some conditions (depending on electrolyte concentration and particle size), flocculation into the secondary minimum may occur. This flocculation is weak and reversible. By increasing the electrolyte concentration, G_{max} decreases till at a given concentration it vanishes and particle coagulation occurs. This is illustrated in Figure 9.3 that shows the variation of G_T with h at various electrolyte concentrations.

Combination of G_{mix} and G_{el} with G_A (the van der Waals attractive energy) gives the total free energy of interaction G_T (assuming there is no contribution from any residual electrostatic interaction), that is,

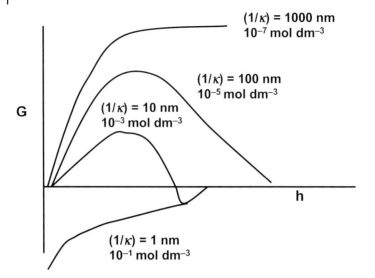

Figure 9.3 Variation of G with h at various electrolyte concentrations.

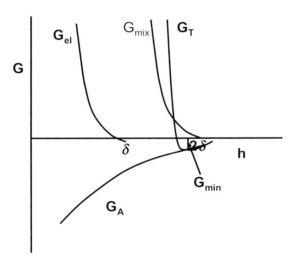

Figure 9.4 Energy–distance curves for sterically stabilized systems.

$$G_T = G_{mix} + G_{el} + G_A \tag{9.13}$$

A schematic representation of the variation of G_{mix}, G_{el}, G_A and G_T with surface–surface separation distance h is shown in Figure 9.4.

G_{mix} increases very sharply with decrease of h, when $h < 2\delta$. G_{el} increases very sharply with decrease of h, when $h < \delta$. G_T versus h shows a minimum, G_{min}, at separation distances comparable to 2δ. When $h < 2\delta$, G_T shows a rapid increase with decrease in h.

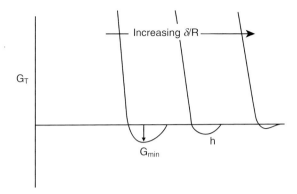

Figure 9.5 Variation of G_T with h at various δ/R values.

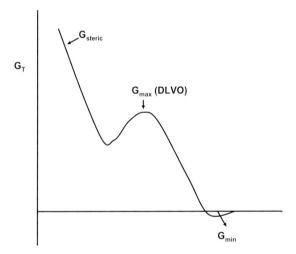

Figure 9.6 Energy–distance curve for electrosteric stabilization.

The depth of the minimum depends on the Hamaker constant A, the particle radius R and adsorbed layer thickness δ. G_{min} increases with increase of A and R. At a given A and R, G_{min} increases with decrease in δ (i.e., with decrease of the molecular weight, M_w, of the stabilizer). This is illustrated in Figure 9.5 which shows the energy–distance curves as a function of δ/R. The larger the value of δ/R, the smaller the value of G_{min}. In this case, the system may approach thermodynamic stability as is the case with nano-dispersions.

Combination of electrostatic repulsion, steric repulsion and van der Waals attraction is referred to as electrosteric stabilization. This is the case, when using a mixture of ionic and nonionic stabilizer or when using polyelectrolytes. In this case, the energy-distance curve has two minima, one shallow maximum (corresponding to the DLVO type) and a rapid increase at small h corresponding to steric repulsion. This is illustrated in Figure 9.6.

9.2
Definition of "Dilute," "Concentrated," and "Solid" Suspensions

It is useful to define the concentration range above a suspension may be considered concentrated. The particle number concentration and volume fraction, ϕ, above which a suspension may be considered concentrated is best defined in terms of the balance between the particle translational motion and interparticle interaction. At one extreme, a suspension may be considered dilute if the thermal motion (Brownian diffusion) of the particles predominate over the imposed interparticle interaction [12, 13]. In this case, the particle translational motion is large and only occasional contacts occur between the particles, that is, the particles do not "see" each other until collision occurs giving a random arrangement of particles. In this case, the particle interactions can be represented by two body collisions. In such "dilute" systems, gravity effects may be neglected and if the particle size range is within the colloid range (1 nm–1 μm) no settling occurs. The properties of the suspension are time independent and, therefore, any time-average quantity such as viscosity or scattering maybe extrapolated to infinite dilution.

As the particle number concentration is increased in a suspension, the volume of space occupied by the particles increases relative to the total volume. Thus, a proportion of the space is excluded in terms of its occupancy by a single particle. Moreover, the particle–particle interaction increases and the forces of interaction between the particles play a dominant role is determining the properties of the system. With further increase in particle number concentration, the interactive contact between the particles increases until a situation is reached where the interaction produces a specific order between the particles, and a highly developed structure is reached. With solid in liquid dispersions, such a highly ordered structure, which is close to the maximum packing fraction ($\varphi = 0.74$ for hexagonally closed packed array of Môn disperse particles) is referred to as "solid" suspension. In such a system, any particle in the system interacts with many neighbors and the vibration amplitude is small relative to the particle size; the properties of the system are essentially time independent [12–14].

In between the random arrangement of particles in "dilute" suspensions and the highly ordered structure of "solid" suspensions, one may easily define "concentrated" suspensions. In this case, the particle interactions occur by many body collisions and the translational motion of the particles is restricted. However, this reduced translational motion is not as great as with "solid" suspensions, that is, the vibrational motion of the particles is large compared with the particle size. A time-dependent system arises in which there will be spatial and temporal correlation.

To understand the property of any dispersion, one must consider the arrangement of the particles in the system: random arrangement with free diffusion, dilute or "vapor-like"; loosely ordered with restricted diffusion, concentrated or "liquid-like"; highly ordered, solid or "crystal-like."

The microstructure of the dispersion may be investigated using small angle x-ray or neutron scattering. Once the microstructure of the system is understood, it is

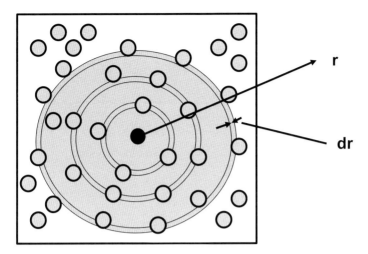

Figure 9.7 Microscopic view (schematic) of the distribution of particles around a central one.

possible to know how the interparticle interaction influences the macrostructure of the system such as its osmotic pressure and rheology.

A convenient method to describe the structure of the suspension is to use the radial distribution function g(r). Consider a system containing N_p particles in a volume V; then the average macroscopic density, ρ_0, is expressed by

$$\rho_0 = \frac{N_p}{V} \quad (9.14)$$

If the container is examined more closely on a microscopic scale, one can obtain the distribution of particles around any reference particle, as is illustrated in Figure 9.7. In the immediate vicinity of the central particle, there is a space in which the particle density is zero. With an increasing distance r from the center of the chosen particle, and circumscribing a ring of thickness dr, we see it contains more particles. As r becomes very large the number of particles within such an annulus will tend to ρ_0. A function must therefore be defined to describe the distribution of particles relative to the central reference particle. This is defined as $\rho(r)$ that varies with r and which describes the distribution of particles. This density function will have two limiting values: $\rho(r) \to 0$ as $r \to 2R$ (where R is the particle radius); $\rho(r) \to \rho_0$ as $r \to \infty$.

The pair distribution function, g(r) can be defined as

$$g(r) = \frac{\rho(r)}{\rho_0} \quad (9.15)$$

It leads to the radial distribution function $4\pi r^2 \rho(r) \cdot g(r)$ has the properties $g(r) \to 0$ as $r \to 2R$ and $g(r) \to 1$ as $r \to \infty$.

g(r) is directly related to the potential $\phi(r)$ of mean force acting between the particles:

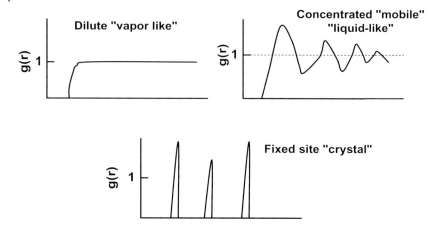

Figure 9.8 Radial distribution function for (a) "dilute"; (b) "concentrated"; (c) "solid" suspensions.

$$g(r) = \exp-\left(\frac{\phi(r)}{kT}\right) \qquad (9.16)$$

$$\phi(r) = V(r) + \psi(r) \qquad (9.17)$$

where $V(r)$ is the simple pair potential and $\psi(r)$ is a perturbation term that takes into account the effect of many body interactions.

For very dilute systems, with particles undergoing Brownian motion, the distribution will be random and only occasional contacts will occur between the particles, that is, there will be only pairwise interaction ($\psi(r) = 0$). In this case, $g(r)$ increases rapidly from the value of zero at $r = 2R$ to its maximum value of unity beyond the first shell (Figure 9.8a). With such dilute systems ("vapor-like"), no structure develops.

For "solid" suspensions, $g(r)$ shows distinct, sharp peaks similar to those observed with atomic and molecular crystals (Figure 9.8c). For "concentrated dispersions" ("liquid-like"), $g(r)$ shows the form represented in Figure 9.8b. This consists of a pronounced first peak followed by a number of oscillatory peaks damping to unity beyond four or five particle diameters. As one proceeds outward from the first shell, the peaks become broader.

The dependence of $g(r)$ on r for colloidal dispersions can be determined by scattering techniques [14]. Figure 9.9a shows the results for polystyrene latex dispersions (particle radius 19 nm) in 10^{-4} mol dm^{-3} at volume fractions of 0.01, 0.04, and 0.13, whereas Figure 9.9b shows the effect of electrolyte concentration (NaCl at 10^{-5}, 10^{-3}, and 5×10^{-3} mol dm^{-3}) at constant volume fraction of 0.04, on the radial distribution function.

The results in Figure 9.9a shows that the $g(r)-r$ curve for the most dilute latex ($\phi = 0.01$) resembles that shown in Figure 9.8 for "dilute" systems. At $\phi = 0.04$ the initial part of the curve becomes much steeper as the particles are closely packed together, and a clear maximum occurs followed by an oscillatory curve resembling

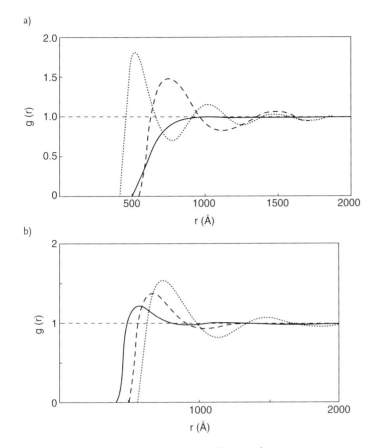

Figure 9.9 (a) g(r) versus r for PS latex in $10^{-4}\,\text{mol}\,\text{dm}^{-3}$ NaCl: ___ $\phi = 0.01$, ---- $\phi = 0.04$, ... $\phi = 0.13$. (b) PS latex, $\phi = 0.04$, ... $C_{\text{NaCl}} = 10^{-5}$, --- $C_{\text{NaCl}} = 10^{-3}$, $C_{\text{NaCl}} = 5 \times 10^{-3}\,\text{mol}\,\text{dm}^{-3}$.

the curve in Figure 9.8 for "concentrated dispersions." At $\phi = 0.13$, the amplitude of the first maximum has increased, indicating more structure; the particles move even closer and are interacting more strongly.

The effect of changing the electrolyte concentration is illustrated in Figure 9.9b that clearly shows that the average distance between the particles is larger the lower the electrolyte concentration. The decrease in the amplitude of the first peak with increasing electrolyte concentration indicates a weakening of the structure.

It is possible, in principle, to relate the microscopic properties of the concentrated dispersion, described earlier, to its macroscopic properties, such its osmotic pressure and the high-frequency shear modulus (rigidity modulus, see Chapter 12). The osmotic pressure is described by

$$\pi = N_p kT - \frac{2\pi N_p^2}{3}\int_0^\infty g(r) r^3 \frac{d\varphi(r)}{dr} dr \tag{9.18}$$

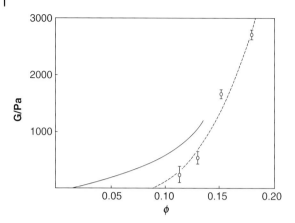

Figure 9.10 Shear modus G versus ϕ for polystyrene latex dispersions in $10^{-4}\,\text{mol\,dm}^{-3}$ NaCl: O, experimental points; _____ calculated based on Eq. (9.19).

The high-frequency shear modulus is given by

$$G_\infty = N_p kT + \frac{2\pi N_p^2}{15} \int_0^\infty g(r) \frac{d}{dr}\left(r^4 \frac{d\varphi(r)}{dr}\right) dr \qquad (9.19)$$

Illustration Figure 9.10 shows the variation of the experimental values of G_∞ with ϕ for polystyrene latex in $10^{-4}\,\text{mol\,dm}^{-3}$ NaCl. The theoretical calculations based on Eq. (9.19) are shown (solid line) on the same figure. The agreement between experimental values of G_∞ is not particularly good and the trends obtained must be regarded as approximate [14].

9.3
States of Suspension on Standing

On standing, concentrated suspensions reach various states (structures) that are determined by (i) magnitude and balance of the various interaction forces, electrostatic repulsion, steric repulsion, and van der Waals attraction; (ii) particle size and shape distribution; (iii) density difference between disperse phase and medium that determines the sedimentation characteristics; (iv) conditions and prehistory of the suspension, for example, agitation that determines the structure of the flocs formed (chain aggregates, compact clusters, etc.); (v) presence of additives, for example, high molecular weight polymers that may cause bridging or depletion flocculation.

An illustration of some of the various states that may be produced is given in Figure 9.11. These states may be described in terms of three different energy–distance curves described earlier. (i) electrostatic, produced for example by the presence of ionogenic groups on the surface of the particles, or adsorption of ionic

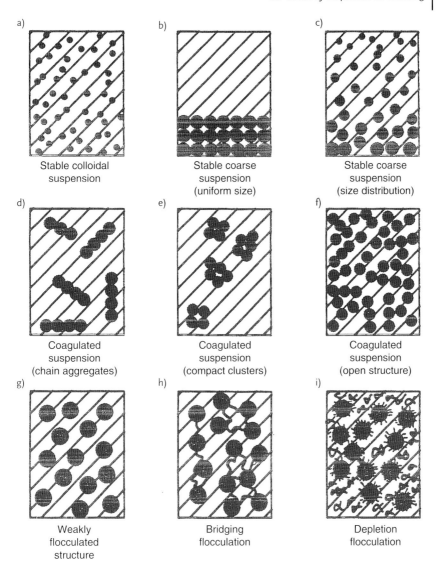

Figure 9.11 States of the suspension.

surfactants. (ii) Steric, produced for example by adsorption of nonionic surfactants or polymers. (iii) Electrostatic + Steric (electrosteric) as for example produced by polyelectrolytes.

A brief description of the various states shown in Figure 9.11 is given in the following [15, 16]:

States (a)–(c) correspond to a suspension that is stable in the colloid sense. The stability is obtained as a result of net repulsion due to the presence of extended double layers (i.e., at low electrolyte concentration), the result of steric repulsion produced adsorption of nonionic surfactants or polymers, or the result of combination of double layer and steric repulsion (electrosteric). State (a) represents the case of a suspension with the small particle size (submicron) whereby the Brownian diffusion overcomes the gravity force producing uniform distribution of the particles in the suspension, that is,

$$kT > (4/3)\pi R^3 \Delta\rho g h \qquad (9.20)$$

where k is the Boltzmann constant, T is the absolute temperature, R is the particle radius, $\Delta\rho$ is the buoyancy (difference in density between the particles and the medium), g is the acceleration due to gravity, and h is the height of the container.

A good example of the earlier case is a latex suspension with the particle size well below $1\,\mu m$ that is stabilized by ionogenic groups, by an ionic surfactant or nonionic surfactant or polymer. This suspension will show no separation on storage for long periods of time.

States (b) and (c) represent the case of suspensions, whereby the particle size range is outside the colloid range (>$1\,\mu m$). In this case, the gravity force exceeds the Brownian diffusion:

$$(4/3)\pi R^3 \Delta\rho g h > kT \qquad (9.21)$$

With state (b), the particles are uniform, and initially they are well dispersed, but with time and the influence of gravity they settle to form hard sediment (technically referred to as "clay" or "cake"). In the sediment, the particles are subjected to a hydrostatic pressure $h\rho g$, where h is the height of the container, ρ is the density of the particles, and g is the acceleration due to gravity. Within the sediment each particle will be acting constantly with many others, and eventually equilibrium is reached where the forces acting between the particles will be balanced by the hydrostatic pressure on the system. The forces acting between the particles will depend on the mechanism used to stabilize the particles, for example electrostatic, steric or electrosteric, the size and shape of the particles, the medium permittivity (dielectric constant), electrolyte concentration, the density of the particles, etc. Many of these factors can be incorporated to give interaction energy in the form of a pair potential for two particles in an infinite medium. The repulsive forces between the particles allow them to move past each other till they reach small distances of separation (that are determined by the location of the repulsive barrier). Due to the small distances between the particles in the sediment, it is very difficult to redisperse the suspension by simple shaking.

With case (c) consisting of a wide distribution of particle sizes, the sediment may contain larger proportions of the larger size particles, but still a hard "clay" is produced. These "clays" are dilatants (i.e., shear thickening) and they can be easily detected by inserting a glass rod in the suspension. Penetration of the glass rod into these hard sediments is very difficult.

States (d)–(f) represent the case for coagulated suspensions that either have a small repulsive energy barrier or its complete absence. State (d) represents the case of coagulation under no stirring conditions in which case chain aggregates are produced that will settle under gravity forming a relatively open structure. State (e) represents the case of coagulation under stirring conditions whereby compact aggregates are produced that will settle faster than the chain aggregates and the sediment produced is more compact. State (f) represents the case of coagulation at high-volume fraction of the particles, φ. In this case, whole particles will form a "one-floc" structure that is formed from chains and cross chains that extend from one wall to the other in the container. Such a coagulated structure may undergo some compression (consolidation) under gravity leaving a clear supernatant liquid layer at the top of the container. This phenomenon is referred to as syneresis.

State (g) represents the case of weak and reversible flocculation. This occurs when the secondary minimum in the energy–distance curve (Figure 9.2) is deep enough to cause flocculation. This can occur at moderate electrolyte concentrations, in particular with larger particles. The same occurs with sterically and electrosterically stabilized suspensions (Figures 9.4 and 9.5). This occurs when the adsorbed layer thickness is not very large, particularly with large particles. The minimum depth required for causing weak flocculation depends on the volume fraction of the suspension. The higher the volume fraction, the lower the minimum depth required for weak flocculation. This can be understood if one considers the free energy of flocculation that consists of two terms, an energy term determined by the depth of the minimum (G_{min}) and an entropy term that is determined by reduction in configurational entropy on aggregation of particles:

$$\Delta G_{flocc} = \Delta H_{flocc} - T\Delta S_{flocc} \tag{9.22}$$

With dilute suspension, the entropy loss on flocculation is larger than with concentrated suspensions. Hence, for flocculation of a dilute suspension, a higher energy minimum is required when compared with the case with concentrated suspensions.

The earlier flocculation is weak and reversible, that is, on shaking the container redispersion of the suspension occurs. On standing, the dispersed particles aggregate to form a weak "gel." This process (referred to as sol gel transformation) leads to reversible time dependence of viscosity (thixotropy). On shearing the suspension, the viscosity decreases and when the shear is removed, the viscosity is recovered. This phenomenon is applied in paints. On application of paint (by a brush or roller), the gel is fluidized, allowing uniform coating of the paint. When shearing is stopped, the paint film recovers its viscosity and this avoids any dripping.

State (h) represents the case whereby the particles are not completely covered by the polymer chains. In this case, simultaneous adsorption of one polymer chain on more than one particle occurs, leading to bridging flocculation. If the polymer adsorption is weak (low adsorption energy per polymer segment), the flocculation could be weak and reversible. In contrast, if the adsorption of the polymer is strong, tough flocs are produced and the flocculation is irreversible. The last phenomenon is used for solid–liquid separation, for example, in water and effluent treatment.

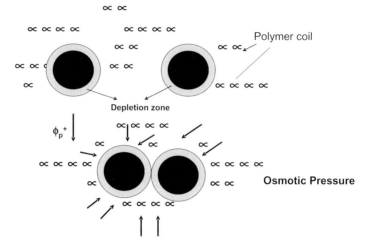

Figure 9.12 Schematic of depletion flocculation.

Case (i) represents a phenomenon, referred to as depletion flocculation, produced by addition of the "free" nonadsorbing polymer [17, 18]. In this case, the polymer coils cannot approach the particles to a distance Δ (that is determined by the radius of gyration of free polymer R_G), since the reduction of entropy on close approach of the polymer coils is not compensated by an adsorption energy. The suspension particles will be surrounded by a depletion zone with thickness Δ. Above a critical volume fraction of the free polymer, φ_p^+, the polymer coils are "squeezed out" from between the particles and the depletion zones begin to interact. The interstices between the particles are now free from polymer coils and hence an osmotic pressure is exerted outside the particle surface (the osmotic pressure outside is higher than in between the particles) resulting in weak flocculation [6, 7]. A schematic representation of depletion flocculation is shown in Figure 9.12.

The magnitude of the depletion attraction free energy, G_{dep}, is proportional to the osmotic pressure of the polymer solution, which in turn is determined by φ_p and molecular weight M. The range of depletion attraction is proportional to the thickness of the depletion zone, Δ, which is roughly equal to the radius of gyration, R_G, of the free polymer. A simple expression for G_{dep} is [7]

$$G_{dep} = \frac{2\pi R \Delta^2}{V_1}(\mu_1 - \mu_1^0)\left(1 + \frac{2\Delta}{R}\right) \qquad (9.23)$$

where V_1 is the molar volume of the solvent, μ_1 is the chemical potential of the solvent in the presence of free polymer with volume fraction φ_p and μ_1^0 is the chemical potential of the solvent in the absence of free polymer. $(\mu_1 - \mu_1^0)$ is proportional to the osmotic pressure of the polymer solution.

References

1 Tadros, T.F. (1987) *Solid/Liquid Dispersions*, Academic Press, London.
2 Tadros, T. (2005) *Applied Surfactants*, Wiley-VCH Verlag GmbH, Weinheim, Germany.
3 Tadros, T.F. (1996) *Adv. Colloid Interface Sci.*, **68**, 97.
4 Gouy, G. (1910) *J. Phys.*, **9** (4), 457; Gouy, G. (1917) *Ann. Phys.*, **7** (9), 129; Chapman, D.L. (1913) *Philos. Mag.*, **25** (6), 475.
5 Stern, O. (1924) *Z. Elektrochem.*, **30**, 508.
6 Bijsterbosch, B.H. (1987) *Solid/Liquid Dispersions* (ed. T.F. Tadros), Academic Press, London.
7 Napper, D.H. (1983) *Polymeric Stabilisation of Colloidal Dispersions*, Academic Press, London.
8 Flory, P.J., and Krigbaum, W.R. (1950) *J. Chem. Phys.*, **18**, 1086.
9 Hamaker, H.C. (1937) *Physica*, **4**, 1058.
10 Deryaguin, B.V., and Landau, L. (1941) *Acta Physicochem. USSR*, **14**, 633.
11 Verwey, E.J.W., and Overbeek, J.T.G. (1948) *Theory of Stability of Lyophobic Colloids*, Elsevier, Amsterdam.
12 Ottewill, R.H. (1982) Chapter 9, in *Concentrated Dispersions* (ed. J.W. Goodwin), Royal Society of Chemistry Publication, N. 43, London.
13 Ottewill, R.H. (1983) *Science and Technology of Polymer Colloids*, vol. II (eds G.W. Poehlein, R.H. Ottewill, and J.W. Goodwin), Martinus Nishof Publishing, Boston, The Hague, p. 503.
14 Ottewill, R.H. (1987) Properties of concentrated suspensions, in *Solid/Liquid Dispersions* (ed. T.F. Taddros), Academic Press, London.
15 Tadros, T.F. (1980) *Adv. Colloid Interface Sci.*, **12**, 141.
16 Tadros, T.F. (1983) *Science and Technology of Polymer Colloids*, vol. II (eds G.W. Poehlein and R.H. Ottewill), Marinus Nishof Publishing, Boston, The Hague.
17 Asakura, A., and Oosawa, F. (1954) *J. Chem. Phys.*, **22**, 1235.
18 Asakura, A., and Oosawa, F. (1958) *J. Polym. Sci.*, **93**, 183.

10
Sedimentation of Suspensions and Prevention of Formation of Dilatant Sediments

Most suspensions undergo separation on standing as a result of the density difference between the particles and the medium, unless the particles are small enough for Brownian motion to overcome gravity [1, 2]. This is illustrated in Figure 10.1 for three cases of suspensions.

Case (a) represents the situation when the Brownian diffusion energy (that is in the region of kT, where k is Boltzmann's constant and T is the absolute temperature) is much larger than the gravitational potential energy (that is equal to $4/3\pi R^3 \Delta\rho g h$, where R is the particle radius, $\Delta\rho$ is the density difference between the particles and medium, g the acceleration due to gravity, and h is the height of the container). Under these conditions, the particles become randomly distributed throughout the whole system, and no separation occurs. This situation may occur with nanosuspensions with radii less than 100 nm, particularly if $\Delta\rho$ is not large, say less than 0.1. In contrast when $4/3\pi R^3 \Delta\rho g h \gg kT$ complete sedimentation occurs as is illustrated in Figure 10.1b with suspensions of uniform particles. In such a case, the repulsive force necessary to ensure colloid stability enables the particles to move past each other to form a compact layer [1, 2]. As a consequence of the dense packing and small spaces between the particles, such compact sediments (that are technically referred to as "clays" or "cakes") are difficult to redisperse. In rheological terms (see Chapter 12), the close packed sediment is shear thickening that is referred to as dilatancy, that is, rapid increase in the viscosity with increase in the shear rate.

The most practical situation is that represented by Figure 10.1c, whereby a concentration gradient of the particles occurs across the container. The concentration of particles C can be related to that at the bottom of the container C_0 by the following equation:

$$C = C_0 \exp\left(-\frac{mgh}{kT}\right) \tag{10.1}$$

where m is the mass of the particles that is given by $(4/3)\pi R^3 \Delta\rho$ (R is the particle radius and $\Delta\rho$ is the density difference between particle and medium), g is the acceleration due to gravity, and h is the height of the container.

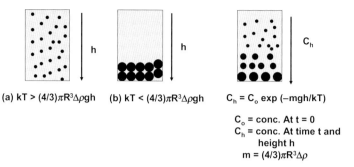

(a) kT > (4/3)πR³Δρgh   (b) kT < (4/3)πR³Δρgh $C_h = C_o \exp(-mgh/kT)$

C_o = conc. At t = 0
C_h = conc. At time t and height h
$m = (4/3)\pi R^3 \Delta\rho$

Figure 10.1 Schematic representation of sedimentation of suspensions.

10.1
Sedimentation Rate of Suspensions

For a very dilute suspension of rigid noninteracting particles, the rate of sedimentation v_0 can be calculated by application of Stokes' law, whereby the hydrodynamic force is balanced by the gravitational force:

$$\text{Hydrodynamic force} = 6\pi \eta R v_0 \tag{10.2}$$

$$\text{Gravity force} = (4/3)\pi R^3 \Delta\rho g \tag{10.3}$$

$$v_0 = \frac{2}{9}\frac{R^2 \Delta\rho g}{\eta} \tag{10.4}$$

where η is the viscosity of the medium (water).

v_0 calculated for three particle sizes (0.1, 1, and 10 μm) for a suspension with density difference $\Delta\rho = 0.2$ is 4.4×10^{-9}, 4.4×10^{-7}, and 4.4×10^{-5} m s^{-1}, respectively. The time needed for complete sedimentation in a 0.1 m container is 250 days, 60 h, and 40 min, respectively.

For moderately concentrated suspensions, $0.2 > \varphi > 0.01$, sedimentation is reduced as a result of hydrodynamic interaction between the particles, which no longer sediment independently of each other [3, 4]. The sedimentation velocity, v, can be related to Stokes' velocity v_0 by the following equation:

$$v = v_0(1 - 6.55\varphi) \tag{10.5}$$

This means that for a suspension with $\varphi = 0.1$, $v = 0.345 v_0$, that is, the rate is reduced by a factor of ~3.

For more concentrated suspensions ($\varphi > 0.2$), the sedimentation velocity becomes a complex function of φ. At $\varphi > 0.4$, one usually enters the hindered settling regime, whereby all the particles sediment at the same rate (independent of the size).

A schematic representation for the variation of v with φ is shown in Figure 10.2, which also shows the variation of relative viscosity with φ. It can be seen that v decreases exponentially with increase in φ and ultimately it approaches zero when φ approaches a critical value ϕ_p (the maximum packing fraction).

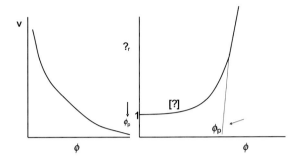

Figure 10.2 Variation of v and η_r with ϕ.

The relative viscosity shows a gradual increase with increase in φ and when $\varphi = \phi$, the relative viscosity approaches infinity.

The maximum packing fraction φ_p can be easily calculated for monodisperse rigid spheres. For hexagonal packing $\varphi_p = 0.74$, whereas for random packing $\varphi_p = 0.64$. The maximum packing fraction increases with polydisperse suspensions. For example, for a bimodal particle size distribution (with a ratio of ~10:1), $\varphi_p > 0.8$.

It is possible to relate the relative sedimentation rate (v/v_0) to the relative viscosity η/η_0:

$$\left(\frac{v}{v_0}\right) = \alpha \left(\frac{\eta_0}{\eta}\right) \tag{10.6}$$

The relative viscosity is related to the volume fraction φ by the Dougherty–Krieger equation [5] for hard spheres:

$$\frac{\eta}{\eta_0} = \left(1 - \frac{\varphi}{\varphi_p}\right)^{-[\eta]\varphi_p} \tag{10.7}$$

where $[\eta]$ is the intrinsic viscosity (= 2.5 for hard spheres).

Combining Eqs. (10.6) and (10.7),

$$\frac{v}{v_0} = \left(1 - \frac{\varphi}{\varphi_p}\right)^{\alpha[\eta]\varphi_p} = \left(1 - \frac{\varphi}{\varphi_p}\right)^{k\varphi_p} \tag{10.8}$$

The above empirical relationship was tested for sedimentation of polystyrene latex suspensions with $R = 1.55\,\mu m$ in $10^{-3}\,\text{mol dm}^{-3}$ NaCl [6]. The results are shown in Figure 10.3.

The circles are the experimental points, whereas the solid line is calculated using Eq. (10.8) with $\varphi_p = 0.58$ and $k = 5.4$.

The sedimentation of particles in non-Newtonian fluids, such as aqueous solutions containing high molecular weight compounds (e.g., hydroxyethyl cellulose or xanthan gum) usually referred to as "thickeners" is not simple since these non-Newtonian solutions are shear thinning with the viscosity decreasing with increase

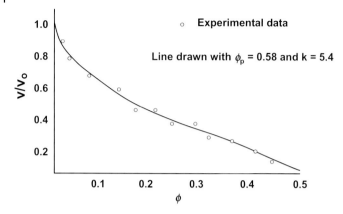

Figure 10.3 Variation of sedimentation rate with volume fraction for polystyrene dispersions.

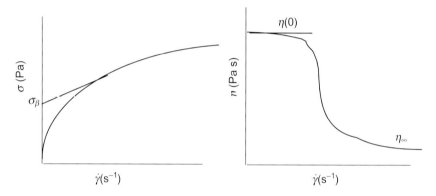

Figure 10.4 Flow behavior of "thickeners."

in shear rate. These solutions show a Newtonian region at low shear rates or shear stresses, usually referred to as the residual or zero shear viscosity $\eta(0)$. This is illustrated in Figure 10.4 that shows the variation of stress σ and viscosity η with shear rate $\dot{\gamma}$.

The viscosity of a polymer solution increases gradually with increase in its concentration and at a critical concentration, C^*, the polymer coils with a radius of gyration R_G and a hydrodynamic radius R_h (that is higher than R_G due to the solvation of the polymer chains) begin to overlap and this shows a rapid increase in viscosity. This is illustrated in Figure 10.5 that shows the variation of $\log \eta$ with $\log C$.

In the first part of the curve $\eta \propto C$, whereas in the second part (above C^*) $\eta \propto C^{3.4}$. A schematic of polymer coil overlap is shown in Figure 10.6 that shows the effect of gradually increasing the polymer concentration. The polymer concentration above C^* is referred to as the semidilute range [7].

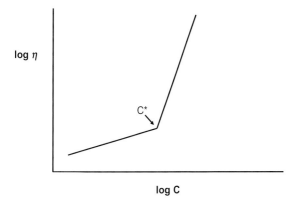

Figure 10.5 Variation of log η with log C.

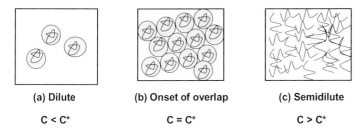

Figure 10.6 Cross-over between dilute and semidilute solutions.

C^* is related to R_G and the polymer molecular weight M by

$$C^* = \frac{3M}{4\pi R_G^3 N_{av}} \qquad (10.9)$$

N_{av} is Avogadro's number. As M increases C^* becomes progressively lower. This shows that to produce physical gels at low concentrations by simple polymer coil overlap, one has to use high molecular weight polymers.

Another method to reduce the polymer concentration at which chain overlap occurs is to use polymers that form extended chains such as xanthan gum that produces conformation in the form of a helical structure with a large axial ratio. These polymers give much higher intrinsic viscosities and they show both rotational and translational diffusion. The relaxation time for the polymer chain is much higher than a corresponding polymer with the same molecular weight but produces random coil conformation.

The above polymers interact at very low concentrations and the overlap concentration can be very low (<0.01%). These polysaccharides are used in many formulations to produce physical gels at very low concentrations thus reducing sedimentation.

The shear stress, σ_p, exerted by a particle (force/area) can be simply calculated [8]:

$$\sigma_p = \frac{(4/3)\pi R^3 \Delta \rho g}{4\pi R^2} = \frac{\Delta \rho R g}{3} \tag{10.10}$$

For a 10 μm radius particle with a density difference $\Delta\rho$ of $0.2\,\mathrm{g\,cm^{-3}}$, the stress is equal to

$$\sigma_p = \frac{0.2 \times 10^3 \times 10 \times 10^{-6} \times 9.8}{3} \approx 6 \times 10^{-3}\,\mathrm{Pa} \tag{10.11}$$

For smaller particles, smaller stresses are exerted.

Thus, to predict sedimentation, one has to measure the viscosity at very low stresses (or shear rates). These measurements can be carried out using a constant stress rheometer (Carrimed, Bohlin, Rheometrics, Haake, or Physica).

Usually one obtains good correlation between the rate of sedimentation v and the residual viscosity $\eta(0)$. Above a certain value of $\eta(0)$, v becomes equal to 0. Clearly to minimize creaming or sedimentation one has to increase $\eta(0)$; an acceptable level for the high shear viscosity η_∞ must be achieved, depending on the application. In some cases, a high $\eta(0)$ may be accompanied by a high η_∞ (that may not be acceptable for application).

As discussed earlier, the stress exerted by the particles is very small, in the region of 10^{-3}–10^{-1} Pa depending on the particle size and the density of particles. Clearly to predict sedimentation, one needs to measure the viscosity at this low stresses [6, 8]. The illustration for solutions of ethyl hydroxy ethyl cellulose (EHEC) is given in Figure 10.7.

The results in Figure 10.7 show that below a certain critical value of the shear stress, the viscous behavior is Newtonian and this critical stress value is in the

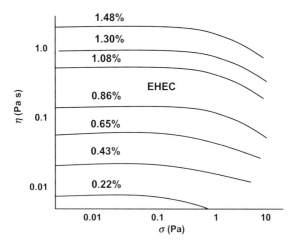

Figure 10.7 Constant stress (creep) measurements for PS latex dispersions as a function of EHEC concentration.

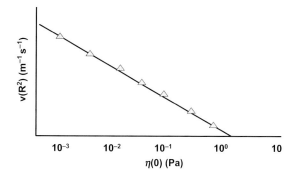

Figure 10.8 Sedimentation rate versus $\eta(0)$.

region of 0.1 Pa. Above this stress, the viscosity decreases with increase in the shear stress, indicating the shear thinning behavior. The plateau viscosity values at low shear stress (<0.1 Pa) give the limiting and residual viscosity $\eta(0)$, that is, the viscosity at near zero shear rate.

The settling rate of a dispersion of polystyrene latex with radius 1.55 μm and at 5% w/v was measured as a function of ethyl(hydroxyethyl) cellulose (EHEC) concentration, using the same range as in Figure 10.7. The settling rate expressed as v/R^2, where R is the particle radius, is plotted versus $\eta(0)$ in Figure 10.8 (on a log–log scale). As is clear, a linear relationship between $\log(v/R^2)$ and $\log \eta(0)$ is obtained, with a slope of −1, over three decades of viscosity, indicating that the rate of settling is proportional to $(\eta(0))^{-1}$.

The maximum shear stress developed by an isolated spherical particle as it settles through a medium of viscosity η is given by the expression [9]:

$$\text{Shear stress} = \frac{3v\eta}{2R} \tag{10.12}$$

For particles at the coarse end of the colloidal range the magnitude of this quantity will be in the range 10^{-2}–10^{-5} Pa. From the data obtained on EHEC solutions and given in Figure 10.7, it can be seen that in this range of shear stresses the solutions behave as Newtonian fluids with zero shear viscosity $\eta(0)$. Hence, isolated spheres should obey Eq. (10.4) with η_0 replaced by $\eta(0)$. Consequently it can be concluded that the rate of sedimentation of the particle is determined by the zero-shear-rate behavior of the medium in which it is suspended. With the present system of polystyrene latex no sedimentation occurred when $\eta(0)$ was greater than 10 Pa.

The situation with more practical dispersions is more complex due to the interaction between the thickener and the particles. Most practical suspensions show some weak flocculation and the "gel" produced between the particles and thickener may undergo some contraction as a result of the gravity force exerted on the whole network. A useful method to describe separation in these concentrated suspensions is to follow the relative sediment volume V_t/V_0 or relative sediment height

h_t/h_0 (where the subscripts t and 0 refers to time t and zero time, respectively) with storage time. For good physical stability the values of V_t/V_0 or h_t/h_0 should be as close as possible to unity (i.e., minimum separation). This can be achieved by balancing the gravitational force exerted by the gel network with the bulk "elastic" modulus of the suspension. The latter is related to the high-frequency modulus G' (see Chapter 12 on rheology).

10.2
Prevention of Sedimentation and Formation of Dilatant Sediments

As mentioned earlier, dilatant sediments are produced with suspensions that are colloidally stable. These dilatant sediments are difficult to redisperse and hence they must be prevented from forming on standing. Several methods may be applied to prevent sedimentation and formation of clays or cakes in a suspension and these are summarized below.

10.2.1
Balance of the Density of the Disperse Phase and Medium

It is clear from Stokes' law that if $\Delta \rho = 0$, $v_0 = 0$. This method can be applied only when the density of the particles is not much larger than that of the medium (e.g., $\Delta \rho \sim 0.1$). By dissolving an inert substance in the continuous phase such as sugar or glycerol one may achieve density matching. However, apart from its limitation to particles with density not much larger than the medium, the method is not very practical since density matching can only occur at one temperature. Liquids usually have large thermal expansion, whereas densities of solids vary comparatively little with temperature.

10.2.2
Reduction of the Particle Size

As mentioned earlier, if R is significantly reduced (to values below $0.1\,\mu\text{m}$), the Brownian diffusion can overcome the gravity force and no sedimentation occurs. This is the principle of formation of nanosuspensions.

10.2.3
Use of High Molecular Weight Thickeners

As discussed above, high molecular weight materials such as hydroxyethyl cellulose or xanthan gum when added above a critical concentration (at which polymer coil overlap occurs) will produce very high viscosity at low stresses or shear rates (usually in excess of several hundred Pascal) and this will prevent sedimentation of the particles. In relatively concentrated suspensions, the situation becomes more complex, since the polymer molecules may lead to flocculation of the suspension, by bridging, depletion (see below), etc. Moreover, the polymer chains at high concentrations tend to interact with each other above a critical concentration C^* (the so-called semidilute region discussed earlier). Such interaction leads to viscoelasticity (see Chapter 12 on rheology), whereby the flow behavior shows an elastic component characterized by an elastic modulus G' (energy elastically stored during deformation) and a viscous component G'' (loss modulus resulting from energy dissipation during flow). The elastic behavior of such relatively concentrated polymer solutions plays a major role in reducing settling and prevention of formation of dilatant clays. A good example of such viscoelastic polymer solution is that of xanthan gum, a high molecular weight polymer (molecular weight in excess of 10^6 Da). This polymer shows viscoelasticity at relatively low concentration (<0.1%) as a result of the interaction of the polymer chains, which are very long. This polymer is very effective in reducing settling of coarse suspensions at low concentrations (in the region of 0.1–0.4% depending on the volume fraction of the suspension).

It should be mentioned that to arrive at the optimum concentration of polymer required to prevent settling and claying of a suspension concentrate, one needs to evaluate the rheological characteristics of the polymer solution, on one hand, and the whole system (suspension and polymer) on the other (see Chapter 12). This will provide the formulator with the necessary information on the interaction of polymer coils with themselves and with the suspended particles. Moreover, one should be careful in applying high molecular-weight materials to prevention of settling, depending on the system. For example, with suspensions used as coatings, such as paints, time effects in flow are very important. In this case, the polymer used for prevention of settling must show reversible time dependence of viscosity (i.e., thixotropy). In other words, the polymer used has to be shear thinning on application to ensure uniform coating, but once the shearing force is removed, the viscosity has to build up quickly in order to prevent undesirable flow. On the other hand, with suspensions that need to be diluted on application, such as agrochemical suspension concentrates, it is necessary to choose a polymer that disperses readily into water, without the need of vigorous agitation. One should also consider the temperature variation of the rheology of the polymer solution. If the rheology undergoes considerable change with temperature, the suspension may clay at high temperatures. One should also consider the aging of the polymer, which may result from chemical or microbiological degradation. This would result in reduction of viscosity with time, and settling or claying may occur on prolonged storage.

10.2.4
Use of "Inert" Fine Particles

Several fine particulate inorganic material produce "gels" when dispersed in aqueous media, for example, sodium montmorillonite or silica. These particulate materials produce three-dimensional structures in the continuous phase as a result of interparticle interaction. For example, sodium montmorillonite (referred to as swellable clays) form gels at low and intermediate electrolyte concentrations.

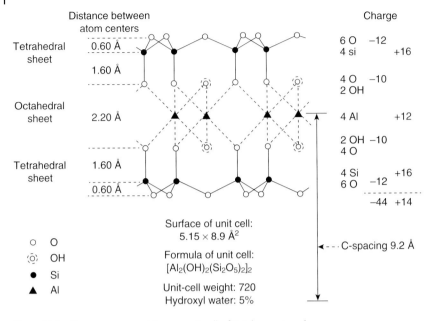

Figure 10.9 Atom arrangement in one unit cell of 2:1 layer mineral.

This can be understood from knowledge of the structure of the clay particles. The latter consist of plate-like particles consisting of a octahedral alumina sheet sandwiched between two tetrahedral silica sheets. This is shown schematically in Figure 10.9, which also shows the change in the spacing of these sheets. In the tetrahedral sheet, tetravalent Si is sometimes replaced by trivalent Al. In the octahedral sheet, there may be replacement of trivalent Al by divalent Mg, Fe, Cr, or Zn. The small size of these atoms allows them to take the place of small Si and Al. This replacement is usually referred to as isomorphic substitution whereby an atom of lower positive valence replaces one of higher valence, resulting in a deficit of positive charge or excess of negative charge. This excess of the negative layer charge is compensated by adsorption at the layer surfaces of cations that are too big to be accommodated in the crystal. In the aqueous solution, the compensation cations on the layer surfaces may be exchanged by other cations in the solution, and hence may be referred to as exchangeable cations. With montmorillonite, the exchangeable cations are located on each side of the layer in the stack, that is, they are present in the external surfaces as well as between the layers. This causes a slight increase of the local spacing from about 9.13 A to about 9.6 A; the difference depends on the nature of the counterion. When montmorillonite clays are placed in contact with water or water vapor the water molecules penetrate between the layers, causing interlayer swelling or (intra)crystalline swelling. This leads to further increase in the basal spacing to 12.5–20 A, depending on the type of clay and cation. This interlayer swelling leads, at most, to doubling of the volume of dry clay where four layers of water are adsorbed. The much larger degree of swelling, which is the driving force for "gel" formation (at low electrolyte concentration), is due to osmotic swelling. It has been suggested that swelling of montmorillonite clays is due to the electrostatic double layers that are produced between the charge layers and cations. This is certainly the case at low electrolyte concentration where the double layer extension (thickness) is large.

As discussed above, the clay particles carry a negative charge as a result of isomorphic substitution of certain electropositive elements by elements of lower valency. The negative charge is compensated by cations, which in the aqueous solution form a diffuse layer, that is, an electric double layer is formed at the clay plate–solution interface. This double layer has a constant charge, which is determined by the type and degree of isomorphic substitution. However, the flat surfaces are not the only surfaces of the plate-like clay particles, they also expose an edge surface. The atomic structure of the edge surfaces is entirely different from that of the flat-layer surfaces. At the edges, the tetrahedral silica sheets and the octahedral alumina sheets are disrupted, and the primary bonds are broken. The situation is analogous to that of the surface of silica and alumina particles in aqueous solution. On such edges, therefore, an electric double layer is created by adsorption of potential determining ions (H^+ and OH^-) and one may, therefore, identify an isoelectric point (i.e.p.) as the point of zero charge (p.z.c.) for these edges. With broken octahedral sheets at the edge, the surface behaves as Al—OH with an i.e.p in the region of pH 7–9. Thus, in most cases the edges become negatively charged above pH 9 and positively charged below pH 9.

van Olphen [10] suggested a mechanism of gel formation of montmorillonite involving interaction of the oppositely charged double layers at the faces and edges of the clay particles. This structure, which is usually referred to as a "card-house" structure, was considered to be the reason for the formation of the voluminous clay gel. However, Norrish suggested that the voluminous gel is the result of the extended double layers, particularly at low electrolyte concentrations. A schematic picture of gel formation produced by double layer expansion and "card-house" structure is shown in Figure 10.10.

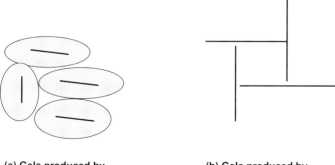

(a) Gels produced by double-layer overlap

(b) Gels produced by edge-to-face association

Figure 10.10 Schematic representation of gel formation in aqueous clay dispersions.

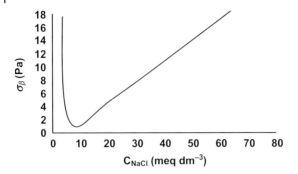

Figure 10.11 Variation of yield value with NaCl concentration for 3.22% sodium montmorillonite dispersions.

Evidence for the above picture was obtained by van Olphen [10] who measured the yield value of 3.22% montmorillonite dispersions as a function of NaCl concentration as shown in Figure 10.11.

When $C = 0$, double layers are extended and gel formation is due to double-layer overlap (Figure 10.11a). First addition of NaCl causes compression of the double layers and hence the yield value decreases very rapidly. At intermediate NaCl concentrations, gel formation occurs as a result of face-to-edge association (house of the card structure) (Figure 10.11b) and the yield value increases very rapidly with increase in NaCl concentration. If the NaCl concentration is increased further, face-to-face association may occur and the yield value decreases (the gel is destroyed).

Finely divided silica such as Aerosil 200 (produced by Degussa) produce gel structures by simple association (by van der Waals attraction) of the particles into chains and cross chains. When incorporated in the continuous phase of a suspension, these gels prevent sedimentation.

10.2.5
Use of Mixtures of Polymers and Finely Divided Particulate Solids

By combining the thickeners such as hydroxyethyl cellulose or xanthan gum with particulate solids such as sodium montmorillonite, a more robust gel structure could be produced. By using such mixtures, the concentration of the polymer can be reduced, thus overcoming the problem of dispersion on dilution (e.g., with many agrochemical suspension concentrates). This gel structure may be less temperature dependent and could be optimized by controlling the ratio of the polymer and the particles. If these combinations of say sodium montmorillonite and a polymer such as hydroxyethyl cellulose, polyvinyl alcohol (PVA) or xanthan gum, are balanced properly, they can provide a "three-dimensional structure," which entraps all the particles and stop settling and formation of dilatant clays. The mechanism of gelation of such combined systems depends to a large extent on the nature of the solid particles, the polymer and the conditions. If the polymer adsorbs on the particle surface (e.g., PVA on sodium montmorillonite or silica) a three-dimensional network may be formed by polymer bridging. Under conditions of incomplete coverage of the particles by the polymer, the latter becomes simultaneously adsorbed on two or more particles. In other words the polymer chains act as "bridges" or "links" between the particles.

10.2.6
Controlled Flocculation ("Self-Structured" Systems)

For systems where the stabilization mechanism is electrostatic in nature, for example, those stabilized by ionic surfactants or polyelectrolytes, the Deryaguin–Landau–Verwey–Overbeek (DLVO) theory [11, 12] predicts the appearance of a secondary attractive minimum at large particle separations. This attractive minimum can reach sufficient values, in particular for large (>1 μm) and asymmetric particles, for weak flocculation to occur. The depth of this minimum does not only depend on particle size, but also on the Hamaker constant, the surface (or zeta) potential and electrolyte concentration and valency. Thus, by careful control of zeta potential and electrolyte concentration, it is possible to arrive at a secondary minimum of sufficient depth for weak flocculation. This results in the formation of a weakly structured "gel" throughout the suspension. This self-structured gel can prevent sedimentation and formation of dilatant clays. As an illustration Figure 10.12 shows energy–distance curves for a suspension stabilized by naphthalene formaldehyde sulfonated condensate (an anionic polyelectrolyte with a modest molecular weight of ~1000) at various NaCl concentrations [13].

It can be seen that at low NaCl concentration, the secondary minimum is absent and hence this suspension will show sedimentation and formation of a dilatant clay. This can be illustrated by measuring the sediment height and no of turns required to redisperse the suspension as is illustrated in Figure 10.13. At 10^{-2} and 5×10^{-2} mol dm^{-3} NaCl, a secondary minimum appears in the energy distance curve that is sufficiently deep for weak flocculation to occur. This is particularly the case with 5×10^{-2} mol dm^{-3} NaCl where the minimum depth reaches ~50 kT. At such concentration, the sediment height increases and the number of turns required for redispersion reaches a small value as is illustrated in Figure 10.13.

The higher the valency of the electrolyte the lower the concentration required to reach a sufficiently deep minimum for weak flocculation to occur. This is clearly shown in Figure 10.13 where the sediment height and number of turns for redispersion are plotted as a function of Na_2SO_4 and $AlCl_3$ concentrations. It can be seen that for Na_2SO_4 a rapid increase in sediment height and decrease of the number of turns for redispersion occurs at $\sim 5 \times 10^{-3}$ mol dm^{-3} whereas for $AlCl_3$ this occurs at $\sim 5 \times 10^{-4}$ mol dm^{-3}. Clearly, the electrolyte concentration (that depends on the electrolyte valency) must be chosen carefully to induce sufficient flocculation to prevent the formation of dilatant sediment, but this concentration should not result in irreversible coagulation (i.e., primary minimum flocculation). With polyelectrolytes, irreversible coagulation is prevented as a result of the contribution of steric interaction (that operates at moderate concentrations of electrolyte).

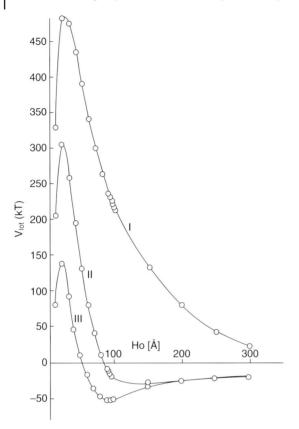

Figure 10.12 Energy–distance curves for electrostatically stabilized suspension at various NaCl concentrations: (i) $10^{-3}\,\mathrm{mol\,dm^{-3}}$; (ii) $10^{-2}\,\mathrm{mol\,dm^{-3}}$; (iii) $5 \times 10^{-2}\,\mathrm{mol\,dm^{-3}}$.

The application of the concept of controlled flocculation to pharmaceutical suspensions has been discussed by various authors [14–17] who demonstrated how flocculation of a sulfaguanidine suspension, by $AlCl_3$, could be used for the preparation of readily dispersible suspensions, which after prolonged storage retain satisfactory physical properties. The work was extended to other types of drug suspensions such as griseofulvin, hydrocortisone, and sulfamerazine [18]. The flocculation observed was interpreted by the DLVO theory; it was suggested that flocculation occurs in a minimum whose depth is restricted owing to steric stabilization by the surfactant film. It should be mentioned that when using $AlCl_3$ in controlling the flocculation, hydrolysable species are produced above pH 4 and these species play a major role in controlling the flocculation.

For systems stabilized with nonionic surfactants or macromolecules, the energy–distance curve also shows a minimum whose depth depends on the particle size and shape, Hamaker constant, and adsorbed layer thickness δ. Thus, for a given particulate system, having a given particle size and shape and Hamaker constant,

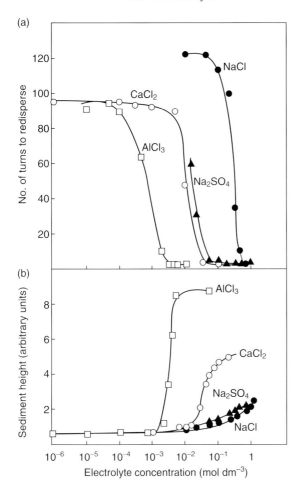

Figure 10.13 Sediment height and redispersion as a function of electrolyte concentration.

the minimum depth can be controlled by varying the adsorbed layer thickness δ. This is illustrated in Figure 10.14, where the energy–distance curves for polystyrene latex particles containing adsorbed polyvinyl alcohol (PVA) layers of various molecular weights, that is, various δ values are shown. These calculations were made using the theory of Hesselink et al. [19] together with the experimentally measured parameters of δ [20]. It is clear from Figure 10.14 that with the high molecular weight PVA fractions δ is large and the energy minimum is too small for flocculation to occur. This is certainly the case with $M = 43\,000$ ($\delta = 19.7$ nm), $M = 28\,000$ ($\delta = 14.0$ nm), and $M = 17\,000$ ($\delta = 9.8$ nm). However with $M = 8000$ ($\delta = 3.3$ nm) an appreciable attraction prevails at a separation distance in the region of 2δ. In this case, a weakly flocculated open structure could be produced for prevention of formation of dilatant sediments. To illustrate this point, dispersions

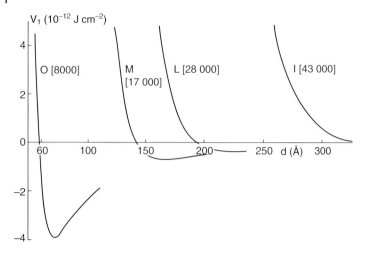

Figure 10.14 Total energy of interaction versus separation distance for polystyrene latex with adsorbed layers of PVA of various thicknesses.

stabilized with PVA of various M were slowly centrifuged at 50 g, and the sediment was freeze dried and examined by electron microscopy. The results are shown in Figure 10.15, which clearly illustrates the closely packed sediment obtained with the high molecular weight PVA fractions and the weakly flocculated open structure obtained with the fraction with low molecular weight.

It should be mentioned that the minimum depth needed to induce flocculation depends on the volume fraction of the suspension. This can be understood if one considers the balance between the interaction energy and entropy terms in the free energy of flocculation, that is,

$$\Delta G_{floc} = \Delta G_h - T\Delta S_h \tag{10.13}$$

where ΔG_h is the interaction energy term (that is negative), which is determined by the depth of the minimum in the energy–distance curve and ΔS_h is the entropy loss on flocculation. On flocculation, configurational entropy is lost and ΔS_h is therefore negative. This means that the term $T\Delta S_h$ is positive (i.e., it opposes flocculation). The condition for flocculation is $\Delta G \leq 0$, and therefore the ΔG_h required for flocculation required for flocculation depends on the magnitude of the $T\Delta S_h$ term. The latter term decreases with increase of the volume fraction ϕ, and hence the higher the ϕ value, the smaller the ΔG_h required for flocculation. This means that with more concentrated suspensions weak flocculation of a sterically stabilized suspension occurs at lower minimum depth.

10.2.7
Depletion Flocculation

As discussed in Chapter 9, addition of free nonadsorbing polymer can produce weak flocculation above a critical volume fraction of the free polymer, ϕ_p^+, which

Figure 10.15 Scanning electron micrographs of polystyrene latex sediments stabilized with PVA fractions: (C) $M = 43\,000$ ($\delta = 19.7$ nm); (L) $M = 28\,000$ ($\delta = 14.0$ nm); (O) $M = 8000$ ($\delta = 3.3$ nm).

depends on its molecular weight and the volume fraction of suspension. This weak flocculation produces a "gel" structure that reduces sedimentation. According to Asakura and Oosawa [21, 22], when two particles approach each other within a distance of separation that is smaller than the diameter of the free polymer coil, exclusion of the polymer molecules from the interstices between the particles takes place, leading to the formation of a polymer-free zone. This is illustrated in Figure 10.16, which shows the situation below and above ϕ_p^+. As a result of this process, an attractive energy, associated with the lower osmotic pressure in the region between the particles, is produced. Fleer et al. [23] derived the following expression for the interaction of hard spheres in the presence of nonadsorbing polymer, that

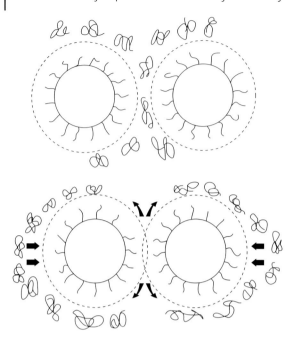

Figure 10.16 Schematic representation of depletion flocculation.

is, the decreases in free energy resulting from the transfer of solvent molecules from the depletion zone to the bulk solution:

$$G_{dep} = \frac{2\pi R}{V_0}(\mu_1 - \mu_1^0)\Delta^2\left(1 + \frac{2\Delta}{3R}\right) \tag{10.14}$$

where V_0 is the molecular volume of the solvent, μ_1 is the solvent chemical potential at ϕ_p and μ_1^0 is the chemical potential of the pure solvent. Since $\mu_1 < \mu_1^0$, $(\mu_1 - \mu_1^0)$ is negative and G_{dep} is negative, resulting in flocculation.

The above phenomenon of flocculation can be applied to the prevention of settling and claying by forming an open structure that, under some conditions, can fill the whole volume of the suspension. Above ϕ_p^+, the suspension becomes weakly flocculated, and the extent of flocculation increases with further increase in the concentration of free nonadsorbing polymer. This is illustrated for a suspension of an agrochemical (ethirimol, a fungicide) that is stabilized by a graft copolymer of poly(methyl methacrylate)/methacrylic acid with poly(ethylene oxide) side chains to which poly(ethylene oxide) (PEO) with $M = 20\,000$, $35\,000$, or $90\,000$ is added for flocculation [24]. Rheological measurements showed that above ϕ_p^+ a rapid increase in the yield value is produced. This is shown in Figure 10.17. Above ϕ_p^+ one would expect significant reduction in formation of dilatant sediments. This is illustrated in Figure 10.18 that shows a plot of sediment height and number of revolutions for redispersion as a function of ϕ_p (for PEO 20 000). It is clear that

10.2 Prevention of Sedimentation and Formation of Dilatant Sediments

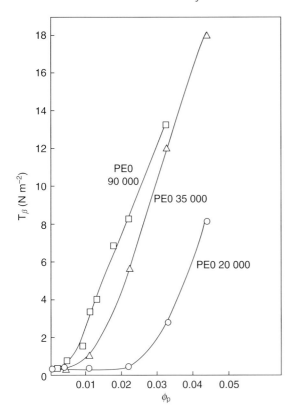

Figure 10.17 Variation of yield value with PEO concentration.

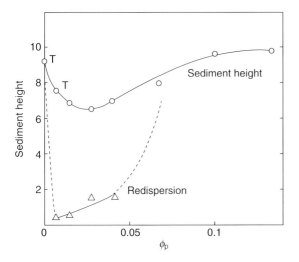

Figure 10.18 Variation of sediment height and redispersion with ϕ_p for PEO 20 000.

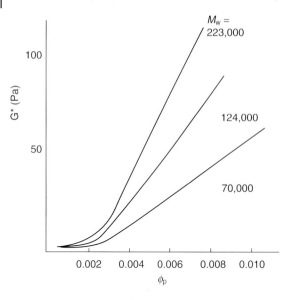

Figure 10.19 Variation of G^* with ϕ_p for HEC with various molecular weights.

addition of PEO results in weak flocculation and redispersion becomes easier. This redispersion is maintained up to $\phi_p = 0.04$ above which it becomes more difficult due to the increase in viscosity.

Another example for the application of depletion flocculation was obtained for the same suspension but using hydroxyethyl cellulose (HEC) with various molecular weights. The weak flocculation was studied using oscillatory measurements. Figure 10.19 shows the variation of the complex modulus G^* with ϕ_p. Above a critical ϕ_p value (that depends on the molecular weight of HEC), G^* increases very rapidly with further increase in ϕ_p. When ϕ_p reaches an optimum concentration, sedimentation is prevented. This is illustrated in Figure 10.20 that shows the sediment volume in 10 cm cylinders as a function ϕ_p for various volume fractions of the suspension ϕ_s. At sufficiently high volume fraction of the suspensions ϕ_s and high volume fraction of free polymer ϕ_p a 100% sediment volume is reached and this is effective in eliminating sedimentation and formation of dilatant sediments.

10.2.8
Use of Liquid Crystalline Phases

Surfactants produce liquid crystalline phases at high concentrations [2]. Three main types of liquid crystals can be identified as illustrated in Figure 10.21: hexagonal phase (sometimes referred to as middle phase), cubic phase, and lamellar (neat phase). All these structures are highly viscous and they also show elastic

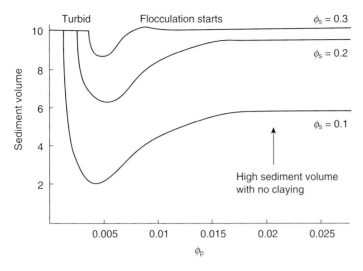

Figure 10.20 Variation of sediment volume with ϕ_p for HEC ($M = 70\,000$).

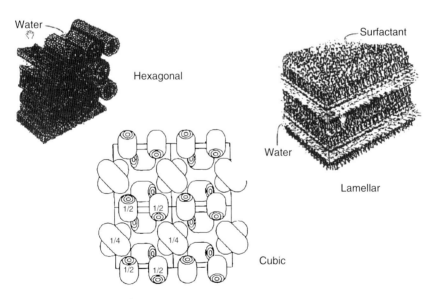

Figure 10.21 Schematic of liquid crystalline phases.

response. If produced in the continuous phase of suspensions, they can eliminate sedimentation of the particles. These liquid crystalline phases are particularly useful for application in liquid detergents that contain high surfactant concentrations. Their presence reduces sedimentation of the coarse builder particles (phosphates and silicates).

References

1 Tadros, T.F. (ed.) (1987) Chapter 11, in *Solid/Liquid Dispersions*, Academic Press, London.
2 Tadros, T. (2005) *Applied Surfactants*, Wiley-VCH Verlag GmbH, Weinheim, Germany.
3 Bachelor, G.K. (1972) *J. Fluid Mech.*, **52**, 245.
4 Bachelor, G.K. (1976) *J. Fluid Mech.*, **79**, 1.
5 Krieger, I.M. (1971) *Adv. Colloid Interface Sci.*, **3**, 45.
6 Buscall, R., Goodwin, J.W., Ottewill, R.H., and Tadros, T.F. (1982) *J. Colloid Interface Sci.*, **85**, 78.
7 de Gennes, P.G. (1979) *Scaling Concepts in Polymer Physics*, Cornell University Press, Ithaca, London.
8 Tadros, T. (2010) *Rheology of Dispersions*, Wiley-VCH Verlag GmbH, Weinheim, Germany.
9 Happel, J., and Brenner, H. (1965) *Low Reynolds Number Hydrodynamics*, Prentice-Hall, London.
10 van Olphen, H. (1963) *Clay Colloid Chemistry*, John Wiley & Sons, Inc., New York.
11 Deryaguin, B.V., and Landau, L. (1941) *Acta Physicochimica. USSR*, **14**, 633.
12 Verwey, E.J.W., and Overbeek, J.T.G. (1948) *Theory of Stability of Lyophobic Colloids*, Elsevier, Amsterdam.
13 Tadros, T.F. (1986) *Colloids Surf.*, **18**, 427.
14 Haines, B.S., and Martin, A.N. (1961) *J. Pharm. Sci.*, **50**, 228, 723, 756.
15 Wilson, R.G., and Canow, B.E. (1963) *J. Pharm. Sci.*, **50**, 757.
16 Mathews, B.A., and Rhodes, C.T. (1986) *J. Pharm. Sci.*, **57**, 557, 569.
17 Jones, R.D.C., Mathews, B.A., and Rhodes, C.T. (1971) *J. Pharm. Sci.*, **59**, 529.
18 Mathews, B.A., and Rhodes, C.T. (1971) *J. Pharm. Sci.*, **59**, 529.
19 Hesselink, F.T., Vrij, A., and Overbeek, J.T.G. (1971) *J. Phys. Chem.*, **75**, 2074.
20 Garvey, M.J., Tadros, T.F., and Vincent, B. (1976) *J. Colloid Interface Sci.*, **55**, 440.
21 Asakura, S., and Oosawa, F. (1954) *J. Phys. Chem.*, **22**, 1255.
22 Asakura, S., and Oosawa, F. (1958) *J. Polym. Sci.*, **33**, 183.
23 Fleer, G.J., Scheutjens, J.H.H.H., and Vincent, B. (1984) *ACS Symp. Ser.*, **246**, 245.
24 Heath, D., Knott, R.D., Knowles, D.A., and Tadros, T.F. (1984) *ACS Symp. Ser.*, **254**, 11.

11
Characterization of Suspensions and Assessment of Their Stability

11.1
Introduction

For full characterization of the properties of suspensions, three main types of investigations are needed: (i) fundamental investigation of the system at a molecular level. This requires investigations of the structure of the solid/liquid interface, namely the structure of the electrical double layer (for charge stabilized suspensions), adsorption of surfactants, polymers, and polyelectrolytes, and conformation of the adsorbed layers (e.g., the adsorbed layer thickness). It is important to know how each of these parameters changes with the conditions, such as temperature, solvency of the medium for the adsorbed layers, and effect of addition of electrolytes. (ii) Investigation of the state of suspension on standing, namely flocculation rates, flocculation points with sterically stabilized systems, spontaneity of dispersion on dilution, and Ostwald ripening or crystal growth. All these phenomena require accurate determination of the particle size distribution as a function of storage time. (iii) Bulk properties of the suspension, that is particularly important for concentrated systems. This requires measurement of the rate of sedimentation and equilibrium sediment height. More quantitative techniques are based on assessment of the rheological properties of the suspension (without disturbing the system, i.e., without its dilution and measurement under conditions of low deformation) and how these are affected by long-term storage. This subject is discussed in detail in Chapter 12.

In this chapter, we will start with a summary of the methods that can be applied to assess the structure of the solid/liquid interface. This is followed by a more detailed section on assessment of sedimentation, flocculation, and Ostwald ripening. For the latter (flocculation and Oswald ripening), one needs to obtain information on the particle size distribution. Several techniques are available for obtaining this information on diluted systems. It is essential to dilute the concentrated suspension with its own dispersion medium in order not to affect the state of the dispersion during examination. The dispersion medium can be obtained by centrifugation of the suspension whereby the supernatant liquid produced at the top (with suspensions) or bottom (with most emulsions) of the centrifuge tube. Care

Dispersion of Powders in Liquids and Stabilization of Suspensions, First Edition. Tharwat F. Tadros.
© 2012 Wiley-VCH Verlag GmbH & Co. KGaA. Published 2012 by Wiley-VCH Verlag GmbH & Co. KGaA.

should be taken while dilution of the concentrated system with its supernatant liquid (i.e., with minimum shear).

11.2
Assessment of the Structure of the Solid/Liquid Interface

11.2.1
Double-Layer Investigation

11.2.1.1 Analytical Determination of Surface Charge

The surface charge on a solid surface can be obtained by determining the adsorption of potential-determining ions at various potentials of the interface [1]. For example, for a silver iodide sol, one determines the adsorption of Ag^+ and I^- ions at various concentrations of Ag^+ and I^- ions in bulk solution. Similarly for an oxide, one determines the adsorption of H^+ and OH^- ions (Γ_{H^+} and Γ_{OH^-}) respectively, as a function of pH of the suspension. In this case, the surface charge density σ_0 is given by

$$\sigma_0 = F(\Gamma_{H^+} - \Gamma_{OH^-}) \tag{11.1}$$

whereas the surface potential ψ_0 is given by the Nernst equation

$$\psi_0 = \frac{RT}{F} \ln \frac{a_{H^+}}{(a_{H^+})_{pzc}} \tag{11.2}$$

where R is the gas constant, T is the absolute temperature, F is the Faraday constant, a_{H+} is the activity of H^+ ions in bulk solution, and $(a_{H+})_{pzc}$ is the value at the point of zero charge.

σ_0 can be directly determined by titration of an oxide suspension in an aqueous solution of indifferent electrolyte (e.g., KCl) using a cell of the type,

$E_1 \,|\, \text{Oxide suspension} \,|\, E_2$

where E_1 is an electrode reversible to H^+ and OH^- ions, such as glass electrode, and E_2 is a reference electrode such as Ag–AgCl. From a knowledge of the amount of H^+ and OH^- ions added and the amount remaining in solution (which can be calculated from a knowledge of the electrical potential of the above cell), Γ_{H+} and Γ_{OH-} can be determined from material balance and knowledge of the surface area of the oxide. The latter can be determined from gas adsorption using the BET method. In order to calculate the absolute values of Γ_{H+} and Γ_{OH-}, one needs to know the pzc. The latter can be located from the common intersection point of the titration curve at various electrolyte concentrations if there is no specific adsorption of ions.

As an illustration of the direct surface charge determination, the results obtained on precipitated silica are shown in Figure 11.1, where σ_0 is plotted versus pH at four different KCl concentrations [2]. There is a common intersection point at pH ~ 3, that is, the pzc, indicating the absence of specific adsorption of K^+ or

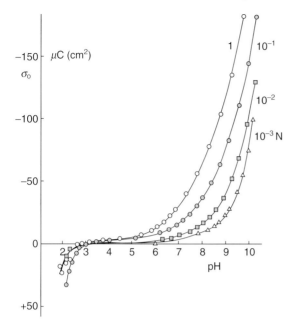

Figure 11.1 σ_0–pH isotherms for precipitated silica at four KCl concentrations.

Cl$^-$ ions. The charge increases progressively with increasing pH reaching very high values at high pH and electrolyte concentrations. However, this high surface charge is not reflected in a high zeta potential, and the silica dispersions are not particularly stable even at high pH. This clearly shows that measurement of the surface charge alone cannot be used as an indication of the stability of the silica dispersion.

11.2.1.2 Electrokinetic and Zeta Potential Measurements

The principles of electrokinetic phenomena and measurement of the zeta potential were discussed in detail in Chapter 5. There are essentially two techniques for measurement of the electrophoretic mobility and zeta potential, namely the ultramicroscopic method and laser velocimetry. As an illustration, Figure 11.2 shows plots of ζ-potential versus pH for goethite FeO(OH) at three electrolyte concentrations [3]. It is clear that below pH 6.9, the particles are positively charged and the ζ-potential increases in magnitude with further decrease in the pH. Above pH 6.9, the particles are negatively charged and the zeta potential increases with further increase in pH. At pH 6.9, the particles are uncharged and this denotes the isoelectric point (iep) of goethite. Below and above the iep, the zeta potential decreases with the increase in electrolyte concentration as a result of double-layer compression. In this case, the stability of goethite is directly correlated with its zeta potential value. The higher the ζ-potential, the more stable the suspension is. Rapid flocculation of goethite suspensions occurs at the iep.

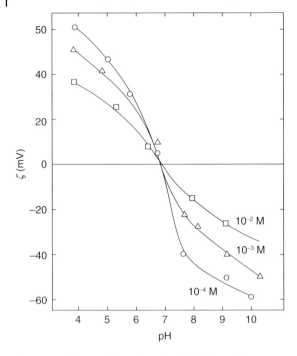

Figure 11.2 ζ-potential as a function of pH for goethite.

11.2.2
Measurement of Surfactant and Polymer Adsorption

As discussed in detail in Chapters 6 and 7, surfactant and polymer adsorption are key to understanding how these molecules affect the stability/flocculation of the suspension. The various techniques that may be applied for obtaining information on surfactant and polymer adsorption have been described before in Chapters 6 and 7. Surfactant (both ionic and nonionic) adsorption is reversible, and the process of adsorption can be described using the Langmuir isotherm [4]. Basically, representative samples of the solid with mass m and surface area A (m^{2}g^{-1}) are equilibrated with surfactant solutions covering various concentrations C_1 (a wide concentration range from values below and above the critical micelle concentration, cmc). The particles are dispersed in the solution by stirring and left to equilibrate (preferably overnight while stirred over rollers) after which the particles are removed by centrifugation and/or filtration (using millipore filters). The concentration in the supernatant solution C_2 is determined using a suitable analytical method. The latter must be sensitive enough for determination of very low surfactant concentration. The surface area of the solid can be determined using gas adsorption and application of the BET equation. Alternatively, the surface area of the "wet" solid (which may be different from that of the dry solid) can be determined using dye adsorption [1].

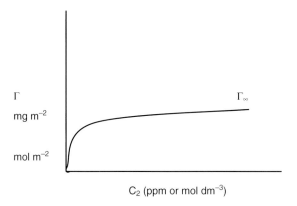

Figure 11.3 Langmuir-type adsorption isotherm.

From the knowledge of C_1 and C_2, and m and A, one can calculate the amount of adsorption Γ (mg m^{-2} or mol m^{-2}) as a function of equilibrium concentration C_2 (ppm or mol dm^{-3}):

$$\Gamma = \frac{C_1 - C_2}{mA} \tag{11.3}$$

With most surfactants, a Langmuir-type isotherm is obtained as is illustrated in Figure 11.3. Γ increases gradually with the increase of C_2 and eventually reaches a plateau value Γ_∞ which corresponds to saturation adsorption.

The results of Figure 11.3 can be fitted to the Langmuir equation

$$\Gamma = \frac{\Gamma_\infty b C_2}{1 + b C_2} \tag{11.4}$$

where b is a constant that is related to the free energy of adsorption ΔG_{ads}:

$$b = \exp(-\Delta G_{ads}/RT) \tag{11.5}$$

A linearized form of the Langmuir equation may be used to obtain Γ_∞ and b as illustrated in Figure 11.4:

$$\frac{1}{\Gamma} = \frac{1}{\Gamma_\infty} + \frac{1}{\Gamma_\infty b C_2} \tag{11.6}$$

A plot of $1/\Gamma$ versus $1/C_2$ gives a straight line (Figure 11.4) with intercept $1/\Gamma_\infty$ and slope $1/\Gamma_\infty b$ from which both Γ_∞ and b can be calculated.

From Γ_∞, the area per surfactant ion or molecule can be calculated:

$$\text{Area/molecule} = \frac{1}{\Gamma_\infty N_{av}}(m^2) = \frac{10^{18}}{\Gamma_\infty N_{av}}(nm^2) \tag{11.7}$$

As discussed in Chapter 6, the area per surfactant ion or molecule gives information on the orientation of surfactant ions or molecules at the interface. This information is relevant for the stability of the suspension. For example, for vertical

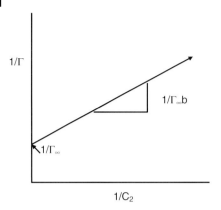

Figure 11.4 Linearized form of the Langmuir equation.

orientation of surfactant ions, for example, dodecyl sulfate anions, which is essential to produce a high surface charge (and hence enhanced electrostatic stability), the area per molecule is determined by the cross-sectional area of the sulfate group which is in the region of $0.4\,\text{nm}^2$. With nonionic surfactants consisting of an alkyl chain and poly(ethylene oxide) (PEO), head group adsorption on a hydrophobic surface is determined by the hydrophobic interaction between the alky chain and the hydrophobic surface. For vertical orientation of a monolayer of surfactant molecules, the area per molecule depends on the size of the PEO chain. The latter is directly related to the number of EO units in the chain. If the area per molecule is smaller than that predicted from the size of the PEO chain, the surfactant molecules may associate on the surface forming bilayers, hemimicelles, etc., as discussed in detail in Chapter 5. This information can be directly related to the stability of the suspension.

The adsorption of polymers is more complex than surfactant adsorption since one must consider the various interactions (chain–surface, chain–solvent, and surface–solvent) as well as the conformation of the polymer chain on the surface [5]. As discussed in Chapter 6, complete information on polymer adsorption may be obtained if one is able to determine the segment density distribution, that is, the segment concentration in all layers parallel to the surface. However, such information is generally unavailable, and therefore one determines three main parameters: the amount of adsorption Γ per unit area, the fraction p of segments in direct contact with the surface (i.e., in trains), and the adsorbed layer thickness δ.

The amount of adsorption Γ can be determined in the same way as for surfactants although in this case the adsorption process may take a long equilibrium time. Most polymers show a high affinity isotherm as is illustrated in Figure 11.5.

This implies that the first added molecules are completely adsorbed, and the isotherm cuts the y axis at $C_2 = 0$. For desorption to occur, the polymer concentration in the supernatant liquid must approach zero, and this implies irreversible adsorption. As discussed in Chapter 6, the magnitude of saturation adsorption

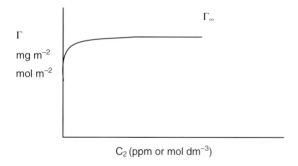

Figure 11.5 High affinity isotherm.

depends on the molecular weight of the polymer, the temperature, and the solvency of the medium for the chains.

The fraction of segments p in trains can be determined using spectroscopic techniques such IR, ESR, and NMR. As discussed in Chapter 6, p depends on surface coverage, polymer molecular weight, and solvency of the medium for the chains.

Several techniques may be applied for determination of the adsorbed layer thickness δ, and these were described in detail in Chapter 6.

11.3
Assessment of Sedimentation of Suspensions

As mentioned in Chapter 10, most suspensions undergo sedimentation on standing due to gravity and the density difference $\Delta\rho$ between the particles and dispersion medium. This is particularly the case when the particle radius exceeds 50 nm and when $\Delta\rho > 0.1$. In this case, the Brownian diffusion cannot overcome the gravity force and sedimentation occurs resulting in increasing the particle concentration from the top to the bottom of the container. As discussed in Chapter 10 to prevent particle sedimentation, "thickeners" (rheology modifiers are added in the continuous phase. The sedimentation of the suspension is characterized by the sedimentation rate, sediment volume, the change of particle size distribution during settling, and the stability of the suspension to sedimentation. Assessment of sedimentation of a suspension depends on the force applied to the particles in the suspension, namely gravitational, centrifugal, and electrophoretic. The sedimentation processes are complex and subject to various errors in sedimentation measurements [6]. A suspension is usually agitated before measuring sedimentation, to ensure an initially homogeneous system of particles in random motion. Vigorous agitation or the use of ultrasonic cavitation must be avoided to prevent any breakdown of aggregates and change of the particle size distribution.

Several physical measurements can be applied to assess sedimentation, and these methods have been described in detail by Kissa [6]. The simplest method is to measure the density of the settling suspension at a known depth using a hydrometer. Unfortunately, this simple method is highly invasive due to the disturbance of the suspension by the hydrometer. A more accurate method is to use sedimentation balances, whereby the sediment accumulated at the base of the sedimentation column is collected and weighed. Manometric methods that use a capillary side arm for measuring the difference between the densities of the pure sedimentation fluid and that of the suspension can also be applied. Several electrical methods can be applied to assess sedimentation. Most suspensions have complex electrical permittivities and may require measurement of both the capacitance and conductivity to determine the solid volume fraction at depth h and time t. This method has the advantage of being noninvasive since the sensing electrodes do not have to be in direct contact with the dispersion. A more convenient method is to use ultrasound probes at various heights from the top to the bottom of the sedimentation tube. The ultrasound velocity and attenuation depend on the volume fraction of the suspension, allowing one to obtain the solid content as a function of height in the sedimentation tube. An alternative optical technique is to measure the back scattering of near infrared at various heights from the sedimentation tube. A commercially available apparatus, namely the Turboscan, can be used for this purpose.

Several other techniques have been designed to monitor sedimentation of suspensions of which photosedimentation, X-ray sedimentation, and laser anemometry are perhaps worth mentioning. The simplest sedimentation test is based on visual observation of settling. The turbidity of the suspension is estimated visually, or the height of the sediment and sediment volume are recorded as a function of time. This visual estimation of sedimentation is only qualitative but is adequate in many practical situations. However, the characterization of suspensions and the determination of particle size distribution require quantitative sedimentation methods. Instrumental techniques have been developed for measuring the turbidity of the suspension as a function of time, either by measuring the turbidity of the bulk suspension or by withdrawing a sample at a given height of the settling suspension. The earlier instruments used for measuring the turbidity of suspensions, called nephelometers, have evolved into instruments with a more sophisticated optical system. Photosedimentometers monitor gravitational particle sedimentation by photoelectric measurement of incident light under steady-state conditions. A horizontal beam of parallel light is projected through a suspension in a sedimentation column to a photocell. Double-beam photosedimentometers using matched photocells, one for the sample and the other for the reference beam, were later developed. A more sophisticated method was later introduced, using linear charge-coupled photodiode array as the image sensor to convert the light intensity attenuated by the particles into an electric signal. The output of each of the photodetectors is handled by a computer independently. Hence, the settling distance between any point in the liquid and the surface of the liquid can be measured accurately without using a mechanical device. As a consequence, the particle measurement is rapid, requiring only about 5 min to determine a particle size distribution. The use of fiberoptics has made it possible to scan the sedimentation column without moving parts or with a fiberoptic probe that is moved inside the sedimentation column.

Laser anemometry, also described as laser Doppler velocity measurement (LVD), is a sensitive technique that can extend the range of photosedimentation methods. It has been applied in a sedimentometer to measure particle sizes as low as 0.5 μm.

X-ray sedimentometers measure X-ray absorption to determine concentration gradients in sedimenting suspensions. The use of X-ray and γ-rays has been proposed as transmittance probes that correlate transmitted radiation with the density of suspension. X-ray transmittance T is directly related to the weight of particles by an exponential relationship, analogous to the Lambert–Beer law governing transmittance of visible radiation:

$$\ln T = -A\varphi_s \tag{11.8}$$

where A is a particle, medium, and equipment constant and ϕ_s is the volume fraction of particles in the suspension.

The concentration of particles remaining in the liquid at various sedimentation depth is determined by using a finely collimated beam of X-rays. The time required for the sedimentation measurement is shortened by continuously changing the effective sedimentation depth. The concentration of particles remaining at various depths is measured as a function of time. The X-ray sedimentometers can be used for particles containing elements with atomic numbers above 15 and, therefore, the method cannot be applied to measure sedimentation of organic pigments.

It should be mentioned that gravitational sedimentation is often too slow, particularly if the particles are small and having a density that is not appreciably higher than that of the medium. Application of a centrifugal force accelerates sedimentation allowing one to obtain results within a reasonable time. However, the data obtained by centrifugation do not always correlate with those resulting from settling under gravity. This is particularly the case with suspensions that are weakly flocculated, where the loose structure may break up on application of a centrifugal force. The interaction between the particles may also change on application of a high gravitational force. This casts doubt on the use of centrifugation as an accelerated test for prediction of sedimentation.

11.4
Assessment of Flocculation and Ostwald Ripening (Crystal Growth)

Assessment of flocculation and Ostwald ripening of a suspension requires measurement of the particle size and shape distribution as a function of time. Several techniques may be applied for this purpose, and these are summarized below [6].

11.4.1
Optical Microscopy

This is by far the most valuable tool for a qualitative or quantitative examination of the suspension. Information on the size, shape, morphology, and aggregation of particles can be conveniently obtained with minimum time required for sample preparation. Since individual particles can be directly observed and their shape examined, optical microscopy is considered as the only absolute method for particle characterization. However, optical microscopy has some limitations: the minimum size that can be detected. The practical lower limit for accurate measurement of particle size is $1.0\,\mu m$, although some detection may be obtained down to $0.3\,\mu m$. Image contrast may not be good enough for observation particularly when using a video camera, which is mostly used for convenience. The contrast can be improved by decreasing the aperture of the iris diaphragm but this reduces the resolution. The contrast of the image depends on the refractive index of the particles relative to that of the medium. Hence, the contrast can be improved by increasing the difference between the refractive index of the particles and the immersion medium. Unfortunately, changing the medium for the suspension is not practical since this may affect the state of the dispersion. Fortunately, water with a refractive index of 1.33 is a suitable medium for most organic particles with a refractive index usually >1.4.

The ultramicroscope by virtue of dark-field illumination extends the useful range of optical microscopy to small particles not visible in a bright-light illumination. Dark-field illumination utilizes a hollow cone of light at a large angle of incidence. The image is formed by light scattered from the particles against a dark background. Particles about 10 times smaller than those visible by bright-light illumination can be detected. However, the image obtained is abnormal and the particle size cannot be accurately measured. For that reason, the electron microscope (see below) has displaced the ultramicroscope, except for dynamic studies by flow ultramicroscopy.

Three main attachments to the optical microscope are possible:

1) **Phase contrast:** this utilizes the difference between the diffracted waves from the main image and the direct light from the light source. The specimen is illuminated with a light cone, and this illumination is within the objective aperture. The light illuminates the specimen and generates zero order and higher orders of diffracted light. The zero-order light beam passes through the objective and a phase plate which is located at the objective back focal plane. The difference between the optical path of the direct light beam and that of the beam diffracted by a particle causes a phase difference. The constructive and destructive interferences result in brightness changes which enhance the contrast. This produces sharp images allowing one to obtain particle-size measurements more accurately. The phase contrast microscope has a plate in the focal plane of the objective back focus. The condenser is equipped instead of a conventional iris diaphragm with a ring matched in its dimension to the phase plate.

2) **Differential interference contrast:** this gives a better contrast than the phase-contrast method. It utilizes a phase difference to improve contrast but the separation and recombination of a light beam into two beams are accom-

plished by prisms. Differential interference contrast generates interference colors, and the contrast effects indicate the refractive index difference between the particle and medium.

3) **Polarized light microscopy:** This illuminates the sample with linearly or circularly polarized light, either in a reflection or transmission mode. One polarizing element, located below the stage of the microscope, converts the illumination to polarized light. The second polarizer is located between the objective and the ocular and is used to detect polarized light. Linearly polarized light cannot pass the second polarizer in a crossed position, unless the plane of polarization has been rotated by the specimen. Various characteristics of the specimen can be determined, including anisotropy, polarization colors, birefringence, polymorphism, etc.

11.4.1.1 Sample Preparation for Optical Microscopy

A drop of the suspension is placed on a glass slide and covered with a cover glass. If the suspension has to be diluted, the dispersion medium (that can be obtained by centrifugation and/or filtration of the suspension) should be used as the diluent in order to avoid aggregation. At low magnifications, the distance between the objective and the sample is usually adequate for manipulating the sample, but at high magnification, the objective may be too close to the sample. An adequate working distance can be obtained, while maintaining high magnification, by using a more powerful eyepiece with a low-power objective. For suspensions encountering Brownian motion (when the particle size is relatively small), microscopic examination of moving particles can become difficult. In this case, one can record the image on a photographic film or video tape or disc (using the computer software).

11.4.1.2 Particle Size Measurements Using Optical Microscopy

The optical microscope can be used to observe dispersed particles and flocs. Particle sizing can be carried out using manual, semiautomatic, or automatic image analysis techniques. In the manual method (which is tedious), the microscope is fitted with a minimum of 10× and 43× achromatic or apochromatic objectives equipped with a high numerical apertures (10×, 15×, and 20×), a mechanical XY stage, a stage micrometer, and a light source. The direct measurement of particle size is aided by a linear scale or globe-and-circle graticules in the ocular. The linear scale is useful mainly for spherical particles, with a relatively narrow particle size distribution. The globe-and-circle graticules are used to compare the projected particle area with a series of circles in the ocular graticule. The size of spherical particles can be expressed by the diameter, but for irregularly shape particles, various statistical diameters are used. One of the difficulties with the evaluation of dispersions by optical microscopy is the quantification of data. The number of particles in at least six different size ranges must be counted to obtain a distribution. This problem can be alleviated by the use of automatic image analysis which can also give an indication on the floc size and its morphology.

11.4.2
Electron Microscopy

Electron microscopy utilizes an electron beam to illuminate the sample. The electrons behave as charged particles, which can be focused by annular electrostatic or electromagnetic fields surrounding the electron beam. Due to the very short wavelength of electrons, the resolving power of an electron microscope exceeds that of an optical microscope by ~200 times. The resolution depends on the accelerating voltage, which determines the wavelength of the electron beam, and magnifications as high as 200 000 can be reached with intense beams but this could damage the sample. Mostly the accelerating voltage is kept below 100–200 kV, and the maximum magnification obtained is below 100 000. The main advantage of electron microscopy is the high resolution, sufficient for resolving details separated by only a fraction of a nanometer. The increased depth of field, usually by about 10 μm or about 10 times of that of an optical microscope, is another important advantage of electron microscopy. Nevertheless, electron microscopy has also some disadvantages such as sample preparation, selection of the area viewed, and interpretation of the data. The main drawback of electron microscopy is the potential risk of altering or damaging the sample that may introduce artifacts and possible aggregation of the particles during sample preparation. The suspension has to be dried or frozen, and the removal of the dispersion medium may alter the distribution of the particles. If the particles do not conduct electricity, the sample has to be coated with a conducting layer, such as gold, carbon, or platinum to avoid negative charging by the electron beam. Two main types of electron microscopes are used: transmission and scanning.

11.4.2.1 Transmission Electron Microscopy (TEM)
TEM displays an image of the specimen on a fluorescent screen, and the image can be recorded on a photographic plate or film. The TEM can be used to examine particles in the range 0.001–5 μm. The sample is deposited on a Formvar (polyvinyl formal) film resting on a grid to prevent charging of the simple. The sample is usually observed as a replica by coating with an electron transparent material (such as gold or graphite). The preparation of the sample for the TEM may alter the state of dispersion and cause aggregation. Freeze fracturing techniques have been developed to avoid some of the alterations of the sample during sample preparation. Freeze fracturing allows the dispersions to be examined without dilution, and replicas can be made of dispersions containing water. It is necessary to have a high cooling rate to avoid the formation of ice crystals.

11.4.2.2 Scanning Electron Microscopy (SEM)
SEM can show particle topography by scanning a very narrowly focused beam across the particle surface. The electron beam is directed normally or obliquely at the surface. The backscattered or secondary electrons are detected in a raster pattern and displayed on a monitor screen. The image provided by secondary electrons exhibit good three-dimensional detail. The backscattered electrons,

reflected from the incoming electron beam, indicate regions of high electron density. Most SEMs are equipped with both types of detectors. The resolution of the SEM depends on the energy of the electron beam which does not exceed 30 kV, and hence the resolution is lower than that obtained by the TEM. A very important advantage of SEM is elemental analysis by energy dispersive X-ray analysis (EDX). If the electron beam impinging on the specimen has sufficient energy to excite atoms on the surface, the sample will emit X-rays. The energy required for X-ray emission is characteristic of a given element, and since the emission is related to the number of atoms present, quantitative determination is possible.

Scanning transmission electron microscopy (STEM) coupled with EDX has been used for the determination of metal particle sizes. Specimens for STEM were prepared by ultrasonically dispersing the sample in methanol, and one drop of the suspension was placed onto a Formvar film supported on a copper grid.

11.4.3
Confocal Laser Scanning Microscopy (CLSM)

CLSM is a very useful technique for identification of suspensions. It uses a variable pinhole aperture or variable width slit to illuminate only the focal plane by the apex of a cone of laser light. Out-of-focus items are dark and do not distract from the contrast of the image. As a result of extreme depth discrimination (optical sectioning), the resolution is considerably improved (up to 40% when compared with optical microscopy). The CLSM technique acquires images by laser scanning or uses computer software to subtract out-of-focus details from the in-focus image. Images are stored as the sample is advanced through the focal plane in elements as small as 50 nm. Three-dimensional images can be constructed to show the shape of the particles.

11.4.4
Scanning Probe Microscopy (SPM)

SPM can measure physical, chemical, and electrical properties of the sample by scanning the particle surface with a tiny sensor of high resolution. Scanning probe microscopes do not measure a force directly; they measure the deflection of a cantilever which is equipped with a tiny stylus (the tip) functioning as the probe. The deflection of the cantilever is monitored by (i) a tunneling current, (ii) laser deflection beam from the backside of the cantilever, (iii) optical interferometry, (iv) laser output controlled by the cantilever used as a mirror in the laser cavity, and (v) change in capacitance. SPM generates a three-dimensional image and allows calibrated measurements in three (x, y, z) coordinates. SPB not only produces a highly magnified image but also provides valuable information on sample characteristics. Unlike EM which requires vacuum for its operation, SPM can be operated under ambient conditions and, with some limitation, in liquid media.

11.4.5
Scanning Tunneling Microscopy (STM)

The STM measures an electric current that flows through a thin insulating layer (vacuum or air) separating two conductive surfaces. The electrons are visualized to "tunnel" through the dielectric and generate a current, I, that depends exponentially on the distance, s, between the tiny tip of the sensor and the electrically conductive surface of the sample. The STM tips are usually prepared by etching a tungsten wire in an NaOH solution until the wire forms a conical tip. Pt/Ir wire has also been used. In the contrast current imaging mode, the probe tip is raster-scanned across the surface and a feedback loop adjusts the height of the tip in order to maintain a constant tunnel current. When the energy of the tunneling current is sufficient to excite luminescence, the tip-surface region emits light and functions as an excitation source of subnanometer dimensions. *In situ* STM has revealed a two-dimensional molecular lamellar arrangement of long-chain alkanes adsorbed on the basal plane of graphite. Thermally induced disordering of adsorbed alkanes was studied by variable temperature STM, and atomic scale resolution of the disordered phase was claimed by studying the quenched high-temperature phase.

11.4.6
Atomic Force Microscopy (AFM)

The AFM allows one to scan the topography of a sample using a very small tip made of silicon nitride. The tip is attached to a cantilever that is characterized by its spring constant, resonance frequency, and a quality factor. The sample rests on a piezoceramic tube which can move the sample horizontally (x,y motion) and vertically (z motion). The displacement of the cantilever is measured by the position of a laser beam reflected from the mirrored surface on the top side of the cantilever. The reflected laser beam is detected by a photodetector. AFM can be operated in either a contact or a noncontact mode. In the contact mode the tip travels in close contact with the surface, whereas in the noncontact mode the tip hovers 5–10 nm above the surface.

11.5
Scattering Techniques

These are by far the most useful methods for characterization of suspensions, and in principle they can give quantitative information on the particle size distribution, floc size, and shape. The only limitation of the methods is the need to use sufficiently dilute samples to avoid interference such as multiple scattering which makes interpretation of the results difficult. However, recently backscattering methods have been designed to allow one to measure the sample without dilution. In principle, one can use any electromagnetic radiation such as light, X-ray, or neutrons but in most industrial labs only light scattering is applied (using lasers).

11.5.1
Light Scattering Techniques

These can be conveniently divided into the following classes: (i) time-average light scattering, static or elastic scattering; (ii) turbidity measurements which can be carried out using a simple spectrophotometer; (iii) light diffraction technique; (iv) dynamic (quasielastic) light scattering that is usually referred to as photon correlation spectroscopy – this is a rapid technique that is very suitable for measuring submicron particles or droplets (nano-size range); and (v) backscattering techniques that is suitable for measuring concentrated samples. Application of any of these methods depends on the information required and availability of the instrument.

11.5.1.1 Time-Average Light Scattering

In this method, the dispersion that is sufficiently diluted to avoid multiple scattering is illuminated by a collimated light (usually laser) beam and the time-average intensity of scattered light is measured as a function of scattering angle θ. Static light scattering is termed elastic scattering. Three regimes can be identified.

Rayleigh Regime Whereby the particle radius R is smaller than $\lambda/20$ (where λ is the wavelength of incident light). The scattering intensity is given by the equation

$$I(Q) = (\text{instrument constant})(\text{material constant}) N V_p^2 \tag{11.9}$$

Q is the scattering vector that depends on the wavelength of light λ used and is given by

$$Q = \left(\frac{4\pi n}{\lambda}\right)\sin\left(\frac{\theta}{2}\right) \tag{11.10}$$

where n is the refractive index of the medium.

The material constant depends on the difference between the refractive index of the particle and that of the medium. N is the number of particles and V_p is the volume of each particle. Assuming that the particles are spherical, one can obtain the average size using Eq. (11.1).

The Rayleigh equation reveals two important relationships: (i) the intensity of scattered light increases with the square of the particle volume and consequently with the sixth power of the radius R. Hence, the scattering from larger particles may dominate the scattering from smaller particles. (ii) The intensity of scattering is inversely proportional to λ^4. Hence, a decrease in the wavelength will substantially increase the scattering intensity.

Rayleigh–Gans–Debye Regime (RGD) $\lambda/20 < R < \lambda$ The RGD regime is more complicated than the Rayleigh regime, and the scattering pattern is no longer symmetrical about the line corresponding to the 90° angle but favors forward scattering ($\theta < 90°$) or backscattering (180° > θ > 90°). Since the preference of forward scattering increases with increasing particle size, the ratio $I_{45°}/I_{135°}$ can indicate the particle size.

Mie Regime $R > \lambda$ The scattering behavior is more complex than the RGD regime, and the intensity exhibits maxima and minima at various scattering angles depending on particle size and refractive index. The Mie theory for light scattering can be used to obtain the particle size distribution using numerical solutions. One can also obtain information on particle shape.

11.5.2
Turbidity Measurements

Turbidity (total light scattering technique) can be used to measure particle size, flocculation, and particle sedimentation. This technique is simple and easy to use; a single or double beam spectrophotometer or a nephelometer can be used.

For nonabsorbing particles, the turbidity τ is given by

$$\tau = (1/L)\ln(I_0/I) \tag{11.11}$$

where L is the path length, I_0 is the intensity of incident beam, and I is the intensity of transmitted beam.

The particle size measurement assumes that the light scattering by a particle is singular and independent of other particles. Any multiple scattering complicates the analysis. According to the Mie theory, the turbidity is related to the particle number N and their cross section πr^2 (where r is the particle radius) by

$$\tau = Q\pi r^2 N \tag{11.12}$$

where Q is the total Mie scattering coefficient. Q depends on the particle size parameter α (which depends on particle diameter and wave length of incident light λ) and the ratio of refractive index of the particles and medium m.

Q depends on the particle size parameter α (which depends on particle diameter and wavelength of incident light) and the ratio of refractive index of the particles and medium. Q depends on α in an oscillatory mode exhibits a series of maxima and minima whose position depends on m. For particles with $R < (1/20) \lambda$, $\alpha < 1$, and it can be calculated using the Rayleigh theory. For $R > \lambda$, Q approaches 2 and between these two extremes, the Mie theory is used. If the particles are not monodisperse (as is the case with most practical systems), the particle size distribution must be taken into account. Using this analysis, one can establish the particle size distribution using numerical solutions.

11.5.3
Light Diffraction Techniques

This is a rapid and nonintrusive technique for determination of particle size distribution in the range 2–300 μm with good accuracy for most practical purposes. Light diffraction gives an average diameter over all particle orientations as ran-

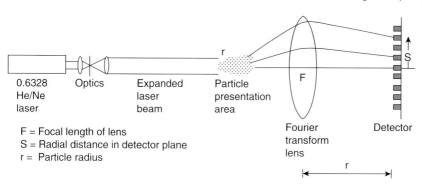

Figure 11.6 Schematic illustration of light diffraction particle sizing system.

domly oriented particles pass the light beam. A collimated and vertically polarized laser beam illuminates particle dispersion and generates a diffraction pattern with the undiffracted beam in the center. The energy distribution of diffracted light is measured by a detector consisting of light sensitive circles separated by isolating circles of equal width. The angle formed by the diffracted light increases with decreasing particle size. The angle-dependent intensity distribution is converted by Fourier optics into a spatial intensity distribution $I(r)$. The spatial intensity distribution is converted into a set of photocurrents, and the particle size distribution is calculated using a computer. Several commercial instruments are available, for example, Malvern Master Sizer (Malvern, UK), Horriba (Japan), and Colter LS Sizer (USA). A schematic illustration of the setup is shown in Figure 11.6.

In accordance with the Fraunhofer theory (which was introduced by Fraunhofer over 100 years ago), the special intensity distribution is given by

$$I(r) = \int_{x_{min}}^{x_{max}} N_{tot} q_0(x) I(r,x) dx \qquad (11.13)$$

where $I(r,x)$ is the radial intensity distribution at radius r for particles of size x, N_{tot} is the total number of particles, and $q_0(x)$ describes the particle size distribution.

The radial intensity distribution $I(r,x)$ is given by

$$I(r,x) = I_0 \left(\frac{\pi x^2}{2f} \right)^2 \left(\frac{J_i(k)}{k} \right)^2 \qquad (11.14)$$

with $k = (\pi x r)/(\lambda f)$

where r is the distance to the center of the disc, λ is the wavelength, f is the focal length, and J_i is the first-order Bessel function.

The Fraunhofer diffraction theory applies to particles whose diameter is considerably larger than the wavelength of illumination. As shown in Figure 11.6, a He/Ne laser is used with $\lambda = 632.8$ nm for particle sizes mainly in the 2–120 μm range.

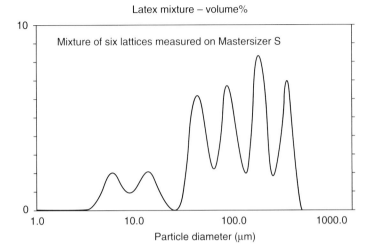

Figure 11.7 Single measurement of a mixture of six standard lattices using the Mastersizer.

In general, the diameter of the sphere-shaped particle should be at least four times the wavelength of the illumination light. The accuracy of particle size distribution determined by light diffraction is not very good if a large fraction of particles with diameter <10 μm is present in the suspension. For small particles (diameter < 10 μm), the Mie theory is more accurate if the necessary optical parameters, such as refractive index of particles and medium and the light absorptivity of the dispersed particles, are known. Most commercial instruments combine light diffraction with forward light scattering to obtain a full particle size distribution covering a wide range of sizes.

As an illustration, Figure 11.7 shows the result of particle sizing using a six component mixture of standard polystyrene lattices (using a Mastersizer).

Most practical suspensions are polydisperse and generate a very complex diffraction pattern. The diffraction pattern of each particle size overlaps with diffraction patterns of other sizes. The particles of different sizes diffract light at different angles, and the energy distribution becomes a very complex pattern. However, manufacturers of light diffraction instruments (such as Malvern, Coulters, and Horriba) developed numerical algorithms relating diffraction patterns to particle size distribution.

Several factors can affect the accuracy of Fraunhofer diffraction: (i) particles smaller than the lower limit of Fraunhofer theory; (ii) nonexistent "ghost" particles in particle size distribution obtained by Fraunhofer diffraction applied to systems containing particles with edges, or a large fraction of small particles (below 10 μm); (iii) computer algorithms that are unknown to the user and vary with the manufacturer software version; (iv) the composition-dependent optical properties of the particles and dispersion medium; and (v) if the density of all particles is not the same, the result may be inaccurate.

11.5.4
Dynamic Light Scattering–Photon Correlation Spectroscopy (PCS)

Dynamic light scattering (DLS) is a method that measures the time-dependent fluctuation of scattered intensity. It is also referred to as quasielastic light scattering (QELS) or photon correlation spectroscopy (PCS). The latter is the most commonly used term for describing the process since most dynamic scattering techniques employ autocorrelation.

PCS is a technique that utilizes the Brownian motion to measure the particle size. As a result of Brownian motion of dispersed particles, the intensity of scattered light undergoes fluctuations that are related to the velocity of the particles. Since larger particles move less rapidly than the smaller ones, the intensity fluctuation (intensity versus time) pattern depends on particle size as is illustrated in Figure 11.8. The velocity of the scatterer is measured in order to obtain the diffusion coefficient.

In a system where the Brownian motion is not interrupted by sedimentation or particle–particle interaction, the movement of particles is random. Hence, the intensity fluctuations observed after a large time interval do not resemble those fluctuations observed initially but represent a random distribution of particles. Consequently, the fluctuations observed at large time delay are not correlated with the initial fluctuation pattern. However, when the time differential between the observations is very small (a nanosecond or a microsecond), both positions of particles are similar and the scattered intensities are correlated. When the time

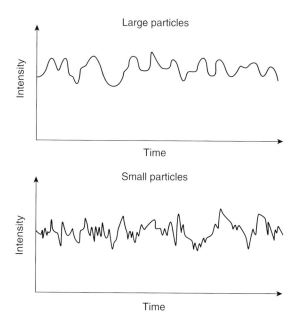

Figure 11.8 Schematic representation of the intensity fluctuation for large and small particles.

interval is increased, the correlation decreases. The decay of correlation is particle size-dependent. The smaller the particles, the faster the decay is.

The fluctuations in scattered light are detected by a photomultiplier and are recorded. The data containing information on the particle motion are processed by a digital correlator. The latter compares the intensity of scattered light at time t, $I(t)$ with the intensity at a very small time interval τ later, $I(t + \tau)$, and it constructs the second-order autocorrelation function $G_2(\tau)$ of the scattered intensity:

$$G_2(\tau) = \langle I(t)I(t+\tau)\rangle \tag{11.15}$$

The experimentally measured intensity autocorrelation function $G_2(\tau)$ depends only on the time interval, τ, and is independent of t, the time when the measurement started.

PCS can be measured in a homodyne where only scattered light is directed to the detector. It can also be measured in heterodyne mode where a reference beam split from the incident beam is superimposed on scattered light. The diverted light beam functions as a reference for the scattered light from each particle.

In the homodyne mode, $G_2(\tau)$ can be related to the normalized field autocorrelation function $g_1(\tau)$ by

$$G_2(\tau) = A + Bg_1^2(\tau) \tag{11.16}$$

where A is the background term designated as the baseline value and B is an instrument-dependent factor. The ratio B/A is regarded as a quality factor of the measurement or the signal-to-noise ratio and expressed sometimes as the % merit.

The field autocorrelation function $g_1(\tau)$ for a monodisperse suspension decays exponentially with τ,

$$g_1(\tau) = \exp(-\Gamma\tau) \tag{11.17}$$

where Γ is the decay constant (s^{-1}).

Substitution of Eq. (11.9) into Eq. (11.8) yields the measured autocorrelation function

$$G_2(\tau) = A + B\exp(-2\Gamma\tau) \tag{11.18}$$

The decay constant Γ is linearly related to the translational diffusion coefficient D_T of the particle:

$$\Gamma = D_T q^2 \tag{11.19}$$

The modulus q of the scattering vector is given by

$$q = \frac{4\pi n}{\lambda_0}\sin\left(\frac{\theta}{2}\right) \tag{11.20}$$

where n is the refractive index of the dispersion medium, θ is the scattering angle, and λ_0 is the wavelength of the incident light in vacuum.

PCS determines the diffusion coefficient, and the particle radius R is obtained using the Stokes–Einstein equation:

$$D = \frac{kT}{6\pi\eta R} \tag{11.21}$$

where k is the Boltzmann constant, T is the absolute temperature, and η is the viscosity of the medium.

The Stokes–Einstein equation is limited to noninteracting, spherical, and rigid spheres. The effect of particle interaction at relatively low particle concentration c can be taken into account by expanding the diffusion coefficient into a power series of concentration:

$$D = D_0(1 + k_D c) \tag{11.22}$$

where D_0 is the diffusion coefficient at infinite dilution and k_D is the virial coefficient that is related to particle interaction. D_0 can be obtained by measuring D at several particle number concentrations and extrapolating to zero concentration.

For polydisperse suspension, the first-order autocorrelation function is an intensity-weighted sum of autocorrelation function of particles contributing to the scattering:

$$g_1(\tau) = \int_0^\infty C(\Gamma)\exp(-\Gamma\tau)\,d\Gamma \tag{11.23}$$

$C(\Gamma)$ represents the distribution of decay rates.

For narrow particle size distribution, the cumulant analysis is usually satisfactory. The cumulant method is based on the assumption that for monodisperse suspensions, $g_1(\tau)$ is mono-exponential. Hence, the log of $g_1(\tau)$ versus τ yields a straight line with a slope equal to Γ:

$$\ln g_1(\tau) = 0.5\ln(B) - \Gamma\tau \tag{11.24}$$

where B is the signal-to-noise ratio.

The cumulant method expands the Laplace transform about an average decay rate

$$\langle\Gamma\rangle = \int_0^\infty \Gamma C(\Gamma)\,d\Gamma \tag{11.25}$$

The exponential in Eq. (11.16) is expanded about an average and integrated term:

$$\ln g_1(\tau) = \langle\Gamma\rangle\tau + (\mu_2\tau^2)/2! - (\mu_3\tau^3)/3! + \cdots \tag{11.26}$$

An average diffusion coefficient is calculated from $\langle\Gamma\rangle$, and the polydispersity (termed the polydispersity index) is indicated by the relative second moment, $\mu_2/\langle\Gamma\rangle^2$. A constrained regulation method (CONTIN) yields several numerical solutions to the particle size distribution, and this is normally included in the software of the PCS machine.

PCS is a rapid, absolute, nondestructive, and rapid method for particle size measurements. It has some limitations. The main disadvantage is the poor resolution of particle size distribution. Also it suffers from the limited size range (absence of any sedimentation) that can be accurately measured. Several instruments are commercially available, for example, by Malvern, Brookhaven, Coulters, etc. The most recent instrument that is convenient to use is HPPS supplied by Malvern (UK), and this allows one to measure the particle size distribution without the need of too much dilution (which may cause some particle dissolution).

11.5.5
Backscattering Techniques

This method is based on the use of fiberoptics, sometimes referred to as fiberoptic dynamic light scattering (FODLS), and it allows one to measure at high particle number concentrations. The FODLS employ either one or two optical fibers. Alternatively, fiber bundles may be used. The exit port of the optical fiber (optode) is immersed in the sample, and the scattered light in the same fiber is detected at a scattering angle of 180° (i.e., backscattering).

The above technique is suitable for online measurements during manufacture of a suspension or emulsion. Several commercial instruments are available, for example, Lesentech (USA).

11.6
Measurement of Rate of Flocculation

Two general techniques may be applied for measuring the rate of flocculation of suspensions, both of which can only be applied for dilute systems. The first method is based on measuring the scattering of light by the particles. For monodisperse particles with a radius that is less than $\lambda/20$ (where λ is the wavelength 0 light), one can apply the Rayleigh equation, whereby the turbidity τ_0 is given by

$$\tau_0 = A' n_0 V_1^2 \tag{11.27}$$

where A' is an optical constant (which is related to the refractive index of the particle and medium and the wavelength of light) and n_0 is the number of particles, each with a volume V_1.

By combining the Rayleigh theory with the Smoluchowski–Fuchs theory of flocculation kinetics [7, 8], one can obtain the following expression for the variation of turbidity with time:

$$\tau = A' n_0 V_1^2 (1 + 2 n_0 k t) \tag{11.28}$$

where k is the rate constant of flocculation.

The second method for obtaining the rate constant of flocculation is by direct particle counting as a function of time. For this purpose, optical microscopy or image analysis may be used, provided that the particle size is within the resolution limit of the microscope. Alternatively, the particle number may be determined using electronic devices such as the Colter counter or the flow ultramicroscope.

The rate constant of flocculation is determined by plotting $1/n$ versus t, where n is the number of particles after time t, that is,

$$\left(\frac{1}{n}\right) = \left(\frac{1}{n_0}\right) + kt \tag{11.29}$$

The rate constant k of slow flocculation is usually related to the rapid rate constant k_0 (the Smoluchowski rate) by the stability ratio W:

$$W = \left(\frac{k}{k_0}\right) \tag{11.30}$$

One usually plots $\log W$ versus $\log C$ (where C is the electrolyte concentration) to obtain the critical coagulation concentration (c.c.c.), which is the point at which $\log W = 0$.

A very useful method for measuring flocculation is to use the single-particle optical method. The particles of the suspension are dispersed in a liquid flow through a narrow uniformly illuminated cell. The suspension is made sufficiently dilute (using the continuous medium) so that particles pass through the cell individually. A particle passing through the light beam illuminating the cell generates an optical pulse detected by a sensor. If the particle size is greater than the wavelength of light (>0.5 μm), the peak height depends on the projected area of the particle. If the particle size is smaller than 0.5 μm, the scattering dominates the response. For particles >1 μm, a light obscuration (also called blockage or extinction) sensor is used. For particles smaller than 1 μm, a light scattering sensor is more sensitive.

The above method can be used to determine the size distribution of aggregating suspensions. The aggregated particles pass individually through the illuminated zone and generate a pulse which is collected at small angle (<3°). At sufficiently small angles, the pulse height is proportional to the square of the number of monomeric units in an aggregate and independent of the aggregate shape or its orientation.

11.7
Measurement of Incipient Flocculation

This can be done for sterically stabilized suspensions, when the medium for the chains becomes a θ-solvent. This occurs, for example, on heating an aqueous suspension stabilized with poly(ethylene oxide) (PEO) or poly(vinyl alcohol) chains. Above a certain temperature (the θ-temperature) that depends on electrolyte concentration, flocculation of the suspension occurs. The temperature at which this occurs is defined as the critical flocculation temperature (CFT).

This process of incipient flocculation can be followed by measuring the turbidity of the suspension as a function of temperature. Above the CFT, the turbidity of the suspension rises very sharply.

For the above purpose, the cell in the spectrophotometer that is used to measure the turbidity is placed in a metal block that is connected to a temperature programming unit (which allows one to increase the temperature rise at a controlled rate).

11.8
Measurement of Crystal Growth (Ostwald Ripening)

Ostwald ripening is the result of the difference in solubility S between small and large particles. The smaller particles have larger solubility than the larger particles. The effect of particle size on solubility is described by the Kelvin equation [9]:,

$$S(r) = S(\infty)\exp\left(\frac{2\sigma V_m}{rRT}\right) \qquad (11.31)$$

where $S(r)$ is the solubility of a particle with radius r, $S(\infty)$ is the solubility of a particle with infinite radius, σ is the solid/liquid interfacial tension, V_m is the molar volume of the disperse phase, R is the gas constant, and T is the absolute temperature.

For two particles with radii r_1 and r_2,

$$\frac{RT}{V_m}\ln\left(\frac{S_1}{S_2}\right) = 2\sigma\left(\frac{1}{r_1} - \frac{1}{r_2}\right) \qquad (11.32)$$

R is the gas constant, T is the absolute temperature, M is the molecular weight, and ρ is the density of the particles.

To obtain a measure of the rate of crystal growth, the particle size distribution of the suspension is followed as a function of time, using either a Coulter counter, a Master sizer, or an optical disc centrifuge. One usually plots the cube of the average radius versus time which gives a straight line from which the rate of crystal growth can be determined (the slope of the linear curve):

$$r^3 = \frac{8}{9}\left[\frac{S(\infty)\sigma V_m D}{\rho RT}\right]t \qquad (11.33)$$

D is the diffusion coefficient of the disperse phase in the continuous phase and ρ is the density of the particles.

11.9
Bulk Properties of Suspensions: Equilibrium Sediment Volume (or Height) and Redispersion

For a "structured" suspension, obtained by "controlled flocculation" or addition of "thickeners" (such as polysaccharides, clays, or oxides), the "flocs" sediment at a rate depending on their size and porosity of the aggregated mass. After this initial sedimentation, compaction and rearrangement of the floc structure occurs, a phenomenon referred to as consolidation.

Normally in sediment volume measurements, one compares the initial volume V_0 (or height H_0) with the ultimately reached value V (or H). A colloidally stable suspension gives a "close-packed" structure with relatively small sediment volume (dilatant sediment referred to as clay). A weakly "flocculated" or "structured" suspension gives a more open sediment and hence a higher sediment volume. Thus by comparing the relative sediment volume V/V_0 or height H/H_0, one can distinguish between a clayed and flocculated suspension.

References

1 Lyklema, J. (1987) Chapter 3, in *Solid/Liquid Dispersions* (ed. T.F. Tadros), Academic Press, London.
2 Tadros, T.F., and Lyklema, J. (1968) *J. Electroanal. Chem.*, **17**, 267.
3 Hunter, R.J. (1981) *Zeta Potential in Colloid Science: Principles and Application*, Academic Press, London.
4 Parfitt, G.D., and Rochester, C.H. (eds) (1983) *Adsorption from Solution at the Solid/Liquid Interface*, Academic Press, London.
5 Fleer, G.J., Cohen-Stuart, M.A., Scheutjens, J.M.H.M., Cosgrove, T., and Vincent, B. (1993) *Polymers at Interfaces*, Chapman and Hall, London.
6 Kissa, E. (1999) *Dispersions, Characterization, Testing and Measurement*, Marcel Dekker, New York.
7 Von Smoluchowski, M. (1916) *Phys. Z.*, **17**, 557, 585.
8 Fuchs, N. (1936) *Z. Phys.*, **89**, 736.
9 Thompson, W. (Lord Kelvin) (1871). *Philos. Mag.*, **42**, 448.

12
Rheological Techniques for Assessment of Stability of Suspensions

12.1
Introduction

Evaluation of the stability/instability of suspensions without any dilution (which can cause significant changes in the structure of the system) requires carefully designed techniques that should cause little disturbance to the structure. The most powerful techniques that can be applied in any industrial laboratory are rheological measurements [1–7]. These measurements provide accurate information on the state of the system such as sedimentation and flocculation. These measurements are also applied for the prediction of the long-term physical stability of the suspension. The various rheological techniques that can be applied and the measurement procedures are listed below.

12.1.1
Steady-State Shear Stress σ–Shear Rate γ Measurements

This requires the use of a shear rate-controlled instrument. The results obtained can be fitted to models to obtain the yield value σ_β and the viscosity η as a function of shear rate. Time effects (thixotropy) can also be investigated.

12.1.2
Constant Stress (Creep) Measurements

A constant is stress is applied on the system, and the strain γ or compliance $J(\gamma/\sigma)$ is followed as a function of time. By measuring creep curves at increasing stress values, one can obtain the residual (zero shear) viscosity $\eta(0)$ and the critical stress σ_{cr} that is the stress above which the structure starts to break down. σ_{cr} is sometimes referred to as the "true" yield value.

Dispersion of Powders in Liquids and Stabilization of Suspensions, First Edition. Tharwat F. Tadros.
© 2012 Wiley-VCH Verlag GmbH & Co. KGaA. Published 2012 by Wiley-VCH Verlag GmbH & Co. KGaA.

12.1.3
Dynamic (Oscillatory) Measurements

A sinusoidal stress or strain with amplitudes σ_0 and γ_0 is applied at a frequency ω (rad s^{-1}), and the stress and strain are simultaneously measured. For a viscoelastic system as is the case with most formulations, the stress and strain amplitudes oscillate with the same frequency but out of phase. The phase angle shift δ is measured from the time shift of the strain and stress sine waves. From σ_0, γ_0, and δ, one can obtain the complex modulus $|G^*|$, the storage modulus G' (the elastic component), and the loss modulus G'' (the viscous component). The results are obtained as a function of strain amplitude and frequency.

12.2
Steady-State Measurements

Most suspensions particularly those with high-volume fraction and/or containing rheology modifiers do not obey Newton's law. This can be clearly shown from plots of shear stress σ versus shear rate as is illustrated in Figure 12.1. Five different flow curves can be identified: (i) Newtonian; (ii) Bingham plastic; (iii) pseudoplastic (shear thinning); (iv) dilatant (shear thickening); (v) yield stress and shear thinning. The variation of viscosity with shear rate for the above five systems is shown in Figure 12.2. Apart from the Newtonian flow (a), all other systems show a change of viscosity with applied shear rate.

12.2.1
Rheological Models for Analysis of Flow Curves

12.2.1.1 Newtonian Systems

$$\sigma = \eta \dot{\gamma} \qquad (12.1)$$

Here, η is independent of the applied shear rate, for example, simple liquids and very dilute dispersions.

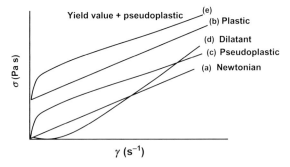

Figure 12.1 Flow curves for various systems.

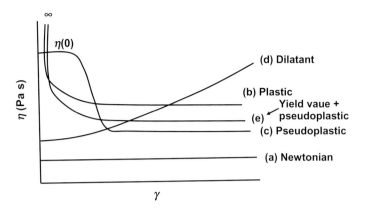

Figure 12.2 Viscosity–shear rate relationship.

12.2.1.2 Bingham Plastic Systems [8]

$$\sigma = \sigma_\beta + \eta_{pl}\dot{\gamma} \qquad (12.2)$$

The system shows a (dynamic) yield stress σ_β that can be obtained by extrapolation to zero shear rate. Clearly at and below σ_β, the viscosity $\eta \rightarrow \infty$. The slope of the linear curve gives the plastic viscosity η_{pl}. Some systems like clay suspensions may show a yield stress above a certain clay concentration.

The Bingham equation describes the shear stress/shear rate behavior of many shear thinning materials at low shear rates. Unfortunately, the value of σ_β obtained depends on the shear rate ranges used for the extrapolation procedure.

12.2.1.3 Pseudoplastic (Shear Thinning) System

In this case, the system does not show a yield value. It shows a limiting viscosity $\eta(0)$ at low shear rates (that is referred to as residual or zero shear viscosity). The flow curve can be fitted to a power law fluid model (Ostwald de Waele):

$$\sigma = k\dot{\gamma}^n \qquad (12.3)$$

where k is the consistency index and n is the shear thinning index, $n < 1$.

By fitting the experimental data to Eq. (12.3), one can obtain k and n. The viscosity at a given shear rate can be calculated:

$$\eta = \frac{\sigma}{\dot{\gamma}} = \frac{k\dot{\gamma}^n}{\dot{\gamma}} = k\dot{\gamma}^{n-1} \qquad (12.4)$$

The power law model (Eq. (12.3)) fits the experimental results for many non-Newtonian systems over two or three decades of shear rate. Thus, this model is more versatile than the Bingham model, although one should be careful in applying this model outside the range of data used to define it. In addition, the power law fluid model fails at high shear rates, whereby the viscosity must ultimately reach a constant value, that is, the value of n should approach unity.

12.2.1.4 Dilatant (Shear Thickening) System

In some cases, the very act of deforming a material can cause rearrangement of its microstructure such that the resistance to flow increases with the increase of shear rate. In other words, the viscosity increases with applied shear rate, and the flow curve can be fitted with the power law, Eq. (12.3), but in this case $n > 1$. The shear thickening regime extends over only about a decade of shear rate. In almost all cases of shear thickening, there is a region of shear thinning at low shear rates.

Several systems can show shear thickening such as wet sand, corn starch dispersed in milk, and some polyvinyl chloride sols. Shear thickening can be illustrated when one walks on wet sand whereby some water is "squeezed out," and the sand appears dry. The deformation applied by one's foot causes rearrangement of the close-packed structure produced by the water motion. This process is accompanied by volume increase (hence the term dilatancy) as a result of "sucking in" of the water. The process amounts to a rapid increase in the viscosity.

12.2.1.5 Herschel–Bulkley General Model [9]

Many systems show a dynamic yield value followed by a shear thinning behavior. The flow curve can be analyzed using the Herscel–Bulkley equation:

$$\sigma = \sigma_\beta + k\dot{\gamma}^n \tag{12.5}$$

When $\sigma_\beta = 0$, Eq. (12.14) reduces to the power fluid model. When $n = 1$, Eq. (12.14) reduces to the Bingham model. When $\sigma_\beta = 0$ and $n = 1$, Eq. (12.15) becomes the Newtonian equation. The Herschel–Bulkley equation fits most flow curves with a good correlation coefficient, and hence it is the most widely used model.

Several other models have been suggested, of which the following is worth mentioning.

12.2.2
The Casson Model [10]

This is a semiempirical linear parameter model that has been applied to fit the flow curves of many paints and printing ink formulations:

$$\sigma^{1/2} = \sigma_C^{1/2} + \eta_C^{1/2}\dot{\gamma}^{1/2} \tag{12.6}$$

Thus a plot of $\sigma^{1/2}$ versus $\dot{\gamma}^{1/2}$ should give a straight line from which σ_C and η_C can be calculated from the intercept and slope of the line. One should be careful in using the Casson equation since straight lines are only obtained from the results above a certain shear rate.

12.2.3
The Cross Equation [11]

This can be used to analyze the flow curve of shear thinning systems that show a limiting viscosity $\eta(0)$ in the low shear rate regime and another limiting viscosity

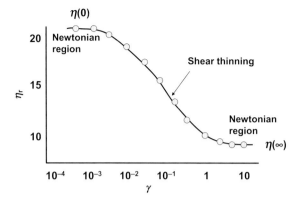

Figure 12.3 Viscosity versus shear rate for shear thinning systems.

$\eta(\infty)$ in the high shear rate regime. These two regimes are separated by a shear thinning behavior as schematically shown in Figure 12.3:

$$\frac{\eta - \eta(\infty)}{\eta(0) - \eta(\infty)} = \frac{1}{1 + K\dot{\gamma}^m} \qquad (12.7)$$

where K is a constant parameter with dimension of time and m is a dimensionless constant.

An equivalent equation to Eq. (12.7) is

$$\frac{\eta_0 - \eta}{\eta - \eta_\infty} = (K\dot{\gamma}^m) \qquad (12.8)$$

12.2.4
Time Effects during Flow Thixotropy and Negative (or anti-) Thixotropy

When a shear rate is applied to a non-Newtonian system, the resulting stress may not be achieved simultaneously. (i) The molecules or particles will undergo spatial rearrangement to follow the applied flow field. (ii) The structure of the system may change: breaking of weak bonds; aligning of irregularly shaped particles; and collision of particles to form aggregates

The above changes are accompanied with the decrease or increase of viscosity with time at any given shear rate. These changes are referred to as thixotropy (if the viscosity decreases with time) or negative thixotropy or antithixotropy (if the viscosity increases with time).

Thixotropy refers to the reversible time-dependent decease of viscosity. When the system is sheared for some time, the viscosity decreases but when the shear is stopped (the system is left to rest), the viscosity of the system is restored. Practical examples for systems that show thixotropy are: paint formulations

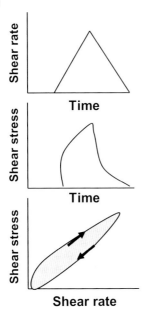

Figure 12.4 Loop test for studying thixotropy.

(sometimes referred to as thixotropic paints); tomato ketchup; and some hand creams and lotions.

Negative thixotropy or antithixotropy: when the system is sheared for sometime, the viscosity increases, but when the shear is stopped (the system is left to rest), the viscosity decreases. A practical example of the above phenomenon is corn starch suspended in milk.

Generally speaking, two methods can be applied to study thixotropy in a suspension. The first and the most commonly used procedure is the loop test whereby the shear rate is increased continuously and linearly in time from zero to some maximum value and then decreased to zero in the same way. This is illustrated in Figure 12.4.

The main problem with the above procedure is the difficulty of interpretation of the results. The nonlinear approach used is not ideal for developing loops because by decoupling the relaxation process from the strain, one does not allow the recovery of the material. However, the loop test gives a qualitative behavior of the suspension thixotropy.

An alternative method for studying thixotropy is to apply a step change test, whereby the suspension is suddenly subjected to a constant high shear rate, and the stress is followed as a function of time whereby the structure breaks down and an equilibrium value is reached. The stress is further followed as a function of time to evaluate the rebuilding of the structure. A schematic representation of this procedure is shown in Figure 12.5.

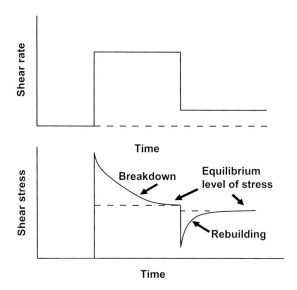

Figure 12.5 Step change for studying thixotropy.

12.3
Constant Stress (Creep) Measurements

A constant stress σ is applied on the system (that may be placed in the gap between two concentric cylinders or a cone and plate geometry), and the strain (relative deformation) γ or compliance $J (= \gamma/\sigma,\ \text{Pa}^{-1})$ is followed as a function of time for a period of t. At $t = t$, the stress is removed and the strain γ or compliance J is followed for another period t [1].

The above procedure is referred to as "creep measurement." From the variation of J with t when the stress is applied and the change of J with t when the stress is removed (in this case J changes sign), one can distinguish between viscous, elastic, and viscoelastic response as is illustrated in Figure 12.6.

Viscous response: in this case, the compliance J shows a linear increase with the increase of time reaching a certain value after time t. When the stress is removed after time t, J remains the same, that is, in this case no creep recovery occurs.

Elastic response: in this case, the compliance J shows a small increase at $t = 0$, and it remains almost constant for the whole period t. When the stress is removed, J changes sign and it reaches 0 after some time t, that is, complete creep recovery occurs in this case.

Viscoelastic response: at $t = 0$, J shows a sudden increase, and this is followed by slower increase for the time applied. When the stress is removed, J changes sign and J shows an exponential decrease with the increase of time (creep recovery) but it does not reach 0 as with the case of an elastic response.

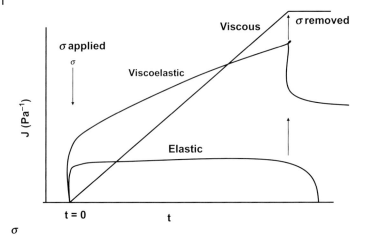

Figure 12.6 Creep curves for viscous, elastic, and viscoelastic response.

12.3.1
Analysis of Creep Curves

12.3.1.1 Viscous Fluid
The linear curve of J versus t gives a slope that is equal to the reciprocal viscosity:

$$J(t) = \frac{\gamma}{\sigma} = \frac{\dot{\gamma}t}{\sigma} = \frac{t}{\eta(0)} \tag{12.9}$$

12.3.1.2 Elastic Solid
The increase of compliance at $t = 0$ (rapid elastic response) $J(t)$ is equal to the reciprocal of the instantaneous modulus $G(0)$:

$$J(t) = \frac{1}{G(0)} \tag{12.10}$$

12.3.2
Viscoelastic Response

12.3.2.1 Viscoelastic Liquid
Figure 12.4 shows the case for a viscoelastic liquid whereby the compliance $J(t)$ is given by two components – an elastic component J_e that is given by the reciprocal of the instantaneous modulus and a viscous component J_v that is given by $t/\eta(0)$:

$$J(t) = \frac{1}{G(0)} + \frac{t}{\eta(0)} \tag{12.11}$$

Figure 12.7 also shows the recovery curve which gives $\sigma_0 J_e^0$, and when this is subtracted from the total compliance gives $\sigma_0 t/\eta(0)$.

Creep is the sum of a constant value $J_e\sigma_0$
(elastic part) and a viscous contribution $\sigma_0 t/\eta_0$

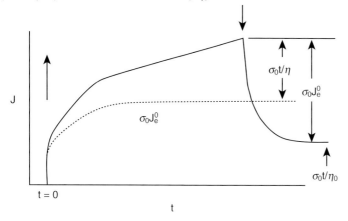

Figure 12.7 Creep curve for a viscoelastic liquid.

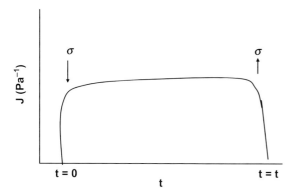

Figure 12.8 Creep curve for a viscoelastic solid.

The driving force for relaxation is spring, and the viscosity controls the rate. The Maxwell relaxation time τ_M is given by

$$\tau_M = \frac{\eta(0)}{G(0)} \tag{12.12}$$

12.3.2.2 Viscoelastic Solid

In this case, complete recovery occurs as illustrated in Figure 12.8. The system is characterized by a Kelvin retardation time τ_k that is also given by the ratio of $\eta(0)/G(0)$.

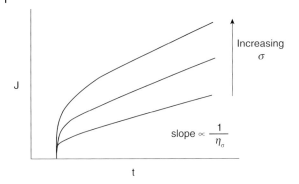

Figure 12.9 Creep curves at increasing applied stress.

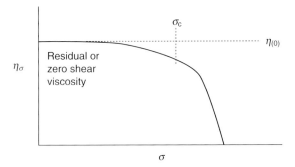

Figure 12.10 Variation of viscosity with applied stress.

12.3.3
Creep Procedure

In creep experiments, one starts with a low applied stress (below the critical stress σ_{cr}, see below) at which the system behaves as a viscoelastic solid with complete recovery as is illustrated in Figure 12.8. The stress is gradually increased, and several creep curves are obtained. Above σ_{cr}, the system behaves as a viscoelastic liquid showing only partial recovery as is illustrated in Figure 12.7. Figure 12.9 shows a schematic representation of the variation of compliance J with time t at increasing σ (above σ_{cr}).

From the slopes of the lines, one can obtain the viscosity η_σ at each applied stress. A plot of η_σ versus σ is shown in Figure 12.10. This shows a limiting viscosity $\eta(0)$ below σ_{cr}, and above σ_{cr}, the viscosity shows a sharp decrease with further increase in σ. $\eta(0)$ is referred to as the residual or zero shear viscosity which is an important parameter for predicting sedimentation. σ_{cr} is the critical stress above which the structure "breaks down." It is sometimes referred to as the "true" yield stress.

12.4
Dynamic (Oscillatory) Measurements [1]

This is the response of the material to an oscillating stress or strain. When a sample is constrained in, say, a cone and plate or concentric cylinder assembly, an oscillating strain at a given frequency ω (rad s^{-1}) ($\omega = 2\nu\pi$, where ν is the frequency in cycles s^{-1} or Hz) can be applied to the sample. After an initial start-up period, a stress develops in response of the applied strain, that is, it oscillates with the same frequency. The change of the sine waves of the stress and strain with time can be analyzed to distinguish between elastic, viscous, and viscoelastic response. Analysis of the resulting sine waves can be used to obtain the various viscoelastic parameters as discussed below.

Three cases can be considered:

Elastic response: whereby the maximum of the stress amplitude is at the same position as the maximum of the strain amplitude (no energy dissipation). In this case, there is no time shift between stress and strain sine waves.

Viscous response: whereby the maximum of the stress is at the point of maximum shear rate (i.e., the inflection point) where there is maximum energy dissipation. In this case, the strain and stress sine waves are shifted by $\omega t = \pi/2$ (referred to as the phase angle shift δ which in this case is 90°)

Viscoelastic response: In this case, the phase angle shift δ is greater than 0 but less than 90°.

12.4.1
Analysis of Oscillatory Response for a Viscoelastic System

Let us consider the case of a viscoelastic system. The sine waves of strain and stress are shown in Figure 12.11. The frequency ω is in rad s^{-1}, and the time shift

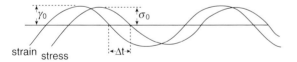

Δt = time shift for sine waves of stress and stain.

$\Delta t \omega = \delta$ phase angle shift

ω = frequency in radian s^{-1}

$\omega = 2\pi\upsilon$

Perfectly elastic solid	$\delta = 0$
Perfectly viscos liquid	$\delta = 90°$
Viscoelastic system	$0 < \delta < 90°$

Figure 12.11 Strain and stress sine waves for a viscoelastic system.

between strain and stress sine waves is Δt. The phase angle shift δ is given by (in dimensionless units of radians)

$$\delta = \omega \Delta t \qquad (12.13)$$

As discussed before,

perfectly elastic solid $\delta = 0$;
perfectly viscous liquid $\delta = 90°$;
viscoelastic system $0 < \delta < 90°$.

The ratio of the maximum stress σ_0 to the maximum strain γ_0 gives the complex modulus $|G^*|$:

$$|G^*| = \frac{\sigma_0}{\gamma_0} \qquad (12.14)$$

The complex modulus can be resolved into G' (the storage or elastic modulus) and G'' (the loss or viscous modulus) using vector analysis and the phase angle shift δ as shown below.

12.4.2
Vector Analysis of the Complex Modulus

$$G' = |G^*|\cos\delta \qquad (12.15)$$
$$G'' = |G^*|\sin\delta \qquad (12.16)$$
$$\tan\delta = \frac{G''}{G'} \qquad (12.17)$$

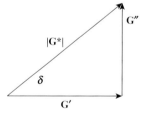

12.4.3
Dynamic Viscosity η'

$$\eta' = \frac{G''}{\omega} \qquad (12.18)$$

12.4.4
Note that $\eta \rightarrow \eta(0)$ as $\omega \rightarrow 0$

Both G' and G'' can be expressed in terms of frequency ω and Maxwell relaxation time τ_m by

$$G'(\omega) = G \frac{(\omega \tau_m)^2}{1 + (\omega \tau_m)^2} \quad (12.19)$$

$$G''(\omega) = G \frac{\omega \tau_m}{1 + (\omega \tau_m)^2} \quad (12.20)$$

In oscillatory techniques, one has to carry two types of experiments:

Strain sweep: the frequency ω is kept constant, and G^*, G', and G'' are measured as a function of strain amplitude.

Frequency sweep: the strain is kept constant (in the linear viscoelastic region), and G^*, G', and G'' are measured as a function of frequency.

12.4.5
Strain Sweep

The frequency is fixed say at 1 Hz (or 6.28 rad s^{-1}), and G^*, G', and G'' are measured as a function of strain amplitude γ_0. This is illustrated in Figure 12.12. G^*, G', and G'' remain constant up to a critical strain γ_{cr}. This is the linear viscoelastic region where the moduli are independent of the applied strain. Above γ_{cr}, G^* and G' start to decrease whereas G'' starts to increase with further increase in γ_0. This is the nonlinear region.

γ_{cr} may be identified with the critical strain above which the structure starts to "break down." It can also be shown that above another critical strain, G'' becomes higher than G'. This is sometimes referred to as the "melting strain" at which the system becomes more viscous than elastic.

Fixed frequency (0.1 or 1 Hz) and follow G^*, G' and G'' with strain amplitude γ_0

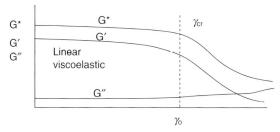

Linear viscoelastic region
G^*, G' and G'' are independent of strain amplitude

γ_{cr} is the critical strain above which system shows non-linear response (break down of structure)

Figure 12.12 Schematic representation of strain sweep.

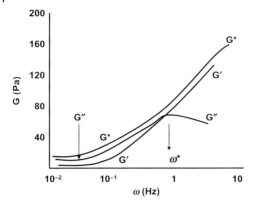

Figure 12.13 Schematic representation of oscillatory measurements for a viscoelastic liquid.

12.4.6
Oscillatory Sweep

The strain γ_0 is fixed in the linear region (taking a mid point, i.e., not a too low strain where the results may show some "noise" and far from γ_{cr}). G^*, G', and G'' are then measured as a function of frequency (a range of 10^{-3}–$10^2\,\mathrm{rad\,s^{-1}}$ may be chosen depending on the instrument and operator patience). Figure 12.13 shows a schematic representation of the variation of G^*, G', and G'' with frequency ω ($\mathrm{rad\,s^{-1}}$) for a viscoelastic system that can be represented by a Maxwell model). One can identify a characteristic frequency ω^* at which $G' = G''$ (the cross-over point) which can be used to obtain the Maxwell relaxation time τ_m:

$$\tau_m = \frac{1}{\omega^*} \tag{12.21}$$

In the low-frequency regime, that is, $\omega < \omega^*$, $G'' > G'$. This corresponds to a long-time experiment (time is reciprocal of frequency), and hence the system can dissipate energy as viscous flow. In the high-frequency regime, that is, $\omega > \omega^*$, $G' > G''$. This corresponds to a short-time experiment where energy dissipation is reduced. At sufficiently high frequency, $G' \gg G''$. At such a high frequency, $G'' \to 0$ and $G' \sim G^*$. The high-frequency modulus $G'(\infty)$ is sometimes referred to as the "rigidity modulus" where the response is mainly elastic.

For a viscoelastic solid, G' does not become zero at low frequency. G'' still shows a maximum at intermediate frequency. This is illustrated in Figure 12.14.

12.4.7
The Cohesive Energy Density E_c

The cohesive energy density, which is an important parameter for identification of the "strength" of the structure in a dispersion can be obtained from the change of G' with γ_0 (see Figure 12.9):

12.4 Dynamic (Oscillatory) Measurements

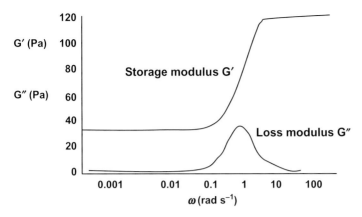

Figure 12.14 Schematic representation for oscillatory measurements for a viscoelastic solid.

$$E_c = \int_0^{\gamma_{cr}} \sigma \, d\gamma \quad (12.22)$$

where σ is the stress in the sample that is given by

$$\sigma = G'\gamma \quad (12.23)$$

$$E_c = \int_0^{\gamma_{cr}} G'\gamma_{cr} \, d\gamma = \frac{1}{2}\gamma_{cr}^2 G' \quad (12.24)$$

Note that E_c is given in $J\,m^{-3}$.

12.4.8
Application of Rheological Techniques for the Assessment and Prediction of the Physical Stability of Suspensions [12]

12.4.8.1 Rheological Techniques for Prediction of Sedimentation and Syneresis

As mentioned in Chapter 11, sedimentation is prevented by addition of "thickeners" that form a "three-dimensional elastic" network in the continuous phase. If the viscosity of the elastic network, at shear stresses (or shear rates) comparable to those exerted by the particles or droplets, exceeds a certain value, then sedimentation is completely eliminated.

The shear stress, σ_p, exerted by a particle (force/area) can be simply calculated,

$$\sigma_p = \frac{(4/3)\pi R^3 \Delta\rho g}{4\pi R^2} = \frac{\Delta\rho R g}{3} \quad (12.25)$$

For a $10\,\mu m$ radius particle with a density difference $\Delta\rho$ of $0.2\,g\,cm^{-3}$, the stress is equal to

$$\sigma_p = \frac{0.2 \times 10^3 \times 10 \times 10^{-6} \times 9.8}{3} \approx 6 \times 10^{-3} \, Pa \quad (12.26)$$

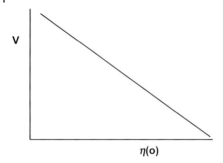

Figure 12.15 Variation of creaming or sedimentation rate with residual viscosity.

For smaller particles, smaller stresses are exerted. Thus, to predict sedimentation, one has to measure the viscosity at very low stresses (or shear rates). These measurements can be carried out using a constant stress rheometer (Carrimed, Bohlin, Rheometrics, Haake, or Physica). Usually one obtains good correlation between the rate of creaming or sedimentation v and the residual viscosity $\eta(0)$. This is illustrated in Figure 12.15. Above a certain value of $\eta(0)$, v becomes equal to 0. Clearly to minimize sedimentation, one has to increase $\eta(0)$; an acceptable level for the high shear viscosity η_∞ must be achieved, depending on the application. In some cases, a high $\eta(0)$ may be accompanied by a high η_∞ (which may not be acceptable for application, for example, if spontaneous dispersion on dilution is required). If this is the case, the formulation chemist should look for an alternative thickener.

Another problem encountered with many dispersions is that of "syneresis," that is, the appearance of a clear liquid film at the top of the container. "Syneresis" occurs with most "flocculated" and/or "structured" (i.e., those containing a thickener in the continuous phase) dispersions. "Syneresis" may be predicted from measurement of the yield value (using steady-state measurements of shear stress as a function of shear rate) as a function of time or using oscillatory techniques (whereby the storage and loss modulus are measured as a function of strain amplitude and frequency of oscillation). The oscillatory measurements are perhaps more useful since to prevent separation the bulk modulus of the system should balance the gravity forces that is given by $h\rho\Delta g$ (where h is the height of the disperse phase, $\Delta\rho$ is the density difference, and g is the acceleration due to gravity). The bulk modulus is related to the storage modulus G'. A more useful predictive test is to calculate the cohesive energy density of the structure E_c that is given by Eq. (12.24).

The separation of a formulation decreases with the increase in E_c. This is illustrated in Figure 12.16 which shows schematically the reduction in % separation with the increase in E_c. The value of E_c that is required to stop complete separation depends on the particle or droplet size distribution, the density difference between the particle or droplet and the medium as well as on the volume fraction ϕ of the dispersion.

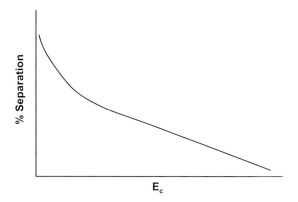

Figure 12.16 Schematic representation of the variation of % separation with E_c.

The correlation of sedimentation with residual (zero shear) viscosity was illustrated in Chapter 11 for model suspensions of aqueous polystyrene latex in the presence of ethylhydroxyethyl cellulose as a thickener.

12.4.8.2 Role of Thickeners

As mentioned above, thickeners reduce creaming or sedimentation by increasing the residual viscosity $\eta(0)$ which must be measured at stresses compared to those exerted by the droplets or particles (mostly less than 0.1 Pa). At such low stresses, $\eta(0)$ increases very rapidly with the increase in "thickener" concentration. This rapid increase is not observed at high stresses, and this illustrates the need for measurement at low stresses (using constant stress or creep measurements). As an illustration, Figure 12.6 shows the variation of η with applied stress σ for ethyl hydroxyethyl cellulose (EHEC), a thickener that is applied in some formulations [13].

As shown in Chapter 11, the limiting residual viscosity increases rapidly with the increase in EHEC concentration. A plot of sedimentation rate for 1.55 μm PS latex particles versus $\eta(0)$ is shown in Figure 12.17, which shows an excellent correlation. In this case, a value of $\eta(0) \geq 10$ Pa s is sufficient for reducing the rate of sedimentation to 0.

12.4.9
Assessment and Prediction of Flocculation Using Rheological Techniques

Steady-state rheological investigations may be used to investigate the state of flocculation of dispersion. Weakly flocculated dispersions usually show thixotropy, and the change of thixotropy with applied time may be used as an indication of the strength of this weak flocculation.

The above methods are only qualitative, and one cannot use the results in a quantitative manner. This is due to the possible breakdown of the structure on transferring the formulation to the rheometer and also during the uncontrolled

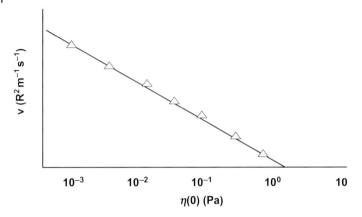

Figure 12.17 Sedimentation rate versus $\eta(0)$.

shear experiment. Better techniques to study flocculation of a formulation are constant stress (creep) or oscillatory measurements. By careful transfer of the sample to the rheometer (with minimum) shear, the structure of the flocculated system may be maintained.

A very important point that must be considered in any rheological measurement is the possibility of "slip" during the measurements. This is particularly the case with highly concentrated dispersions, whereby the flocculated system may form a "plug" in the gap of the platens leaving a thin liquid film at the walls of the concentric cylinder or cone-and-plate geometry. This behavior is caused by some "syneresis" of the formulation in the gap of the concentric cylinder or cone and plate. To reduce "slip," one should use roughened walls for the platens. A vane rheometer may also be used.

Steady-state shear stress-shear rate measurements are by far the most commonly used method in many industrial laboratories. Basically the dispersion is stored at various temperatures, and the yield value σ_β and plastic viscosity η_{pl} are measured at various intervals of time. Any flocculation in the formulation should be accompanied by an increase in σ_β and η_{pl}. A rapid technique to study the effect of temperature changes on the flocculation of a formulation is to carry out temperature sweep experiments, running the samples from 5° to 50°C. The trend in the variation of σ_β and η_{pl} with temperature can quickly give an indication on the temperature range at which a dispersion remains stable (during that temperature range, σ_β and η_{pl} remain constant).

If Ostwald ripening occurs simultaneously, σ_β and η_{pl} may change in a complex manner with storage time. Ostwald ripening results in a shift of the particle size distribution to higher diameters. This has the effect of reducing σ_β and η_{pl}. If flocculation occurs simultaneously (having the effect of increasing these rheological parameters), the net effect may be an increase or decrease of the rheological parameters. This trend depends on the extent of flocculation relative to Ostwald ripening and/or coalescence. Therefore, following σ_β and η_{pl} with storage time requires knowledge of Ostwald ripening. Only in the absence of these latter breakdown processes, one can use rheological measurements as a guide of assessment of flocculation.

Constant stress (creep) experiments are more sensitive for following flocculation. As mentioned before, a constant stress σ is applied on the system and the compliance J (Pa^{-1}) is plotted as a function of time. These experiments are repeated several times increasing the stress from the smallest possible value (that can be applied by the instrument), increasing the stress in small increments. A set of creep curves are produced at various applied stresses. From the slope of the linear portion of the creep curve (after the system reaches a steady state), the viscosity at each applied stress, η_σ, is calculated. A plot of η_σ versus σ allows one to obtain the limiting (or zero shear) viscosity $\eta(0)$ and the critical stress σ_{cr} (which may be identified with the "true" yield stress of the system. The values of $\eta(0)$ and σ_{cr} may be used to assess the flocculation of the dispersion on storage. If flocculation occurs on storage (without any Ostwald ripening), the values of $\eta(0)$ and σ_{cr} may show a gradual increase with the increase of storage time. As discussed in the previous section (on steady-state measurements), the trend becomes complicated if Ostwald ripening occurs simultaneously (both have the effect of reducing $\eta(0)$ and σ_{cr}).

The above measurements should be supplemented by particle size distribution measurements of the diluted dispersion (making sure that no flocs are present after dilution) to assess the extent of Ostwald ripening. Another complication may arise from the nature of the flocculation. If the latter occurs in an irregular way (producing strong and tight flocs), $\eta(0)$ may increase, while σ_{cr} may show some decrease and this complicates the analysis of the results. In spite of these complications, constant stress measurements may provide valuable information on the state of the dispersion on storage.

Carrying out creep experiments and ensuring that a steady state is reached can be time consuming. One usually carries out a stress sweep experiment, whereby the stress is gradually increased (within a predetermined time period to ensure that one is not too far from reaching the steady state), and plots of η_σ versus σ are established. These experiments are carried out at various storage times (say every two weeks) and temperatures. From the change of $\eta(0)$ and σ_{cr} with storage time and temperature, one may obtain information on the degree and the rate of flocculation of the system. Clearly interpretation of the rheological results requires expert knowledge of rheology and measurement of the particle size distribution as a function of time.

One main problem in carrying the above experiments is sample preparation. When a flocculated dispersion is removed from the container, care should be taken not to cause much disturbance to that structure (minimum shear should be applied on transferring the formulation to the rheometer). It is also advisable to use separate containers for assessment of the flocculation; a relatively large sample is prepared, and this is then transferred to a number of separate containers. Each sample is used separately at a given storage time and temperature. One should be careful in transferring the sample to the rheometer. If any separation occurs in the formulation, the sample is gently mixed by placing it on a roller. It is advisable to use as minimum shear as possible when transferring the sample from the container to the rheometer (the sample is preferably transferred using a "spoon" or by simple pouring from the container). The experiment should be carried out without an initial preshear.

Another rheological technique for assessment of flocculation is oscillatory measurement. As mentioned above, one carries out two sets of experiments.

12.4.9.1 Strain Sweep Measurements

In this case, the oscillation is fixed (say at 1 Hz), and the viscoelastic parameters are measured as a function of strain amplitude. G^*, G', and G'' remain virtually constant up to a critical strain value, γ_{cr}. This region is the linear viscoelastic region. Above γ_{cr}, G^* and G' starts to fall, whereas G'' starts to increase; this is the nonlinear region. The value of γ_{cr} may be identified with the minimum strain above which the "structure" of the dispersion starts to break down (for example, breakdown of flocs into smaller units and/or breakdown of a "structuring" agent).

From γ_{cr} and G', one can obtain the cohesive energy E_c (J m^{-3}) of the flocculated structure using Eq. (12.24). E_c may be used in a quantitative manner as a measure of the extent and strength of the flocculated structure in a dispersion. The higher the value of E_c, the more flocculated the structure is. Clearly E_c depends on the volume fraction of the dispersion as well as the particle size distribution (which determines the number of contact points in a floc). Therefore, for quantitative comparison between various systems, one has to make sure that the volume fraction of the disperse particles is the same, and the dispersions have very similar particle size distribution. E_c also depends on the strength of the flocculated structure, that is, the energy of attraction between the droplets. This depends on whether the flocculation is in the primary or secondary minimum. Flocculation in the primary minimum is associated with a large attractive energy, and this leads to higher values of E_c when compared with the values obtained for secondary minimum flocculation (weak flocculation). For a weakly flocculated dispersion, such as the case with secondary minimum flocculation of an electrostatically stabilized system, the deeper the secondary minimum, the higher the value of E_c (at any given volume fraction and particle size distribution of the dispersion). With a sterically stabilized dispersion, weak flocculation can also occur when the thickness of the adsorbed layer decreases. Again the value of E_c can be used as a measure of the flocculation – the higher the value of E_c, the stronger the flocculation. If incipient flocculation occurs (on reducing the solvency of the medium for the change to worse than θ-condition), a much deeper minimum is observed and this is accompanied by a much larger increase in E_c.

To apply the above analysis, one must have an independent method for assessing the nature of the flocculation. Rheology is a bulk property that can give information on the interparticle interaction (whether repulsive or attractive) and to apply it in a quantitative manner, one must know the nature of these interaction forces. However, rheology can be used in a qualitative manner to follow the change of the formulation on storage. Providing the system does not undergo any Ostwald ripening, the change of the moduli with time and in particular the change of the linear viscoelastic region may be used as an indication of flocculation. Strong flocculation is usually accompanied by a rapid increase in G', and this may be accompanied by a decrease in the critical strain above which the "structure" breaks down. This may be used as an indication of formation of "irregular" and tight flocs which become sensitive to the applied strain. The floc structure will entrap a large amount of the continuous phase, and this leads to an apparent increase in the volume fraction of the dispersion and hence an increase in G'.

12.4.9.2 Oscillatory Sweep Measurements

In this case, the strain amplitude is kept constant in the linear viscoelastic region (one usually takes a point far from γ_{cr} but not too low, that is, in the midpoint of the linear viscoelastic region), and measurements are carried out as a function of frequency. Both G^* and G' increase with the increase in frequency and ultimately above a certain frequency, they reach a limiting value and show little dependence on frequency. G'' is higher than G' in the low-frequency regime; it also increases with the increase in frequency, and at a certain characteristic frequency ω^* (that depends on the system), it becomes equal to G' (usually referred to as the cross-over point), after which it reaches a maximum and then shows a reduction with further increase in frequency.

From ω^*, one can calculate the relaxation time τ of the system:

$$\tau = \frac{1}{\omega^*} \tag{12.27}$$

The relaxation time may be used as a guide for the state of the dispersion. For a colloidally stable dispersion (at a given particle size distribution), τ increases with the increase of the volume fraction of the disperse phase, ϕ. In other words, the cross-over point shifts to lower frequency with the increase in ϕ. For a given dispersion, τ increases with the increase in flocculation provided that the particle size distribution remains the same (i.e., no Ostwald ripening).

The value of G' also increases with the increase in flocculation, since aggregation of particles usually result in liquid entrapment, and the effective volume fraction of the dispersion shows an apparent increase. With flocculation, the net attraction between the particles also increases, and this results in an increase in G'. The latter is determined by the number of contacts between the particles and the strength of each contact (which is determined by the attractive energy).

It should be mentioned that in practice, one may not obtain the full curve, due to the frequency limit of the instrument, and also measurement at low frequency is time consuming. Usually one obtains part of the frequency dependence of G' and G''. In most cases, one has a more elastic than viscous system.

Most disperse systems used in practice are weakly flocculated, and they also contain "thickeners" or "structuring" agents to reduce sedimentation and to acquire the right rheological characteristics for application, for example, in hand creams and lotions. The exact values of G' and G'' required depend on the system and its application. In most cases, a compromise has to be made between

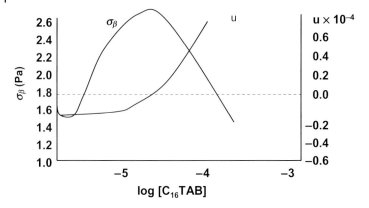

Figure 12.18 Variation of yield value σ_β and electrophoretic mobility u with C_{16}TAB concentration.

acquiring the right rheological characteristics for application and the optimum rheological parameters for long-term physical stability. Application of rheological measurements to achieve these conditions requires a great deal of skill and understanding of the factors that affect rheology.

12.4.10
Examples of Application of Rheology for Assessment and Prediction of Flocculation

12.4.10.1 Flocculation and Restabilization of Clays Using Cationic Surfactants

Hunter and Nicol [14] studied the flocculation and restabilization of kaolinite suspensions using rheology and zeta potential measurements. Figure 12.18 shows plots of the yield value σ_β and electrophoretic mobility as a function of cetyl trimethyl ammonium bromide (CTAB) concentration at pH = 9. σ_β increases with the increase in CTAB concentration, reaching a maximum at the point where the mobility reaches zero (the isoelectric point, i.e.p., of the clay) and then decreases with further increase in CTAB concentration. This trend can be explained on the basis of flocculation and restabilization of the clay suspension.

The initial addition of CTAB causes reduction in the negative surface charge of the clay (by adsorption of CTA^+ on the negative sites of the clay). This is accompanied by reduction in the negative mobility of the clay. When complete neutralization of the clay particles occurs (at the i.e.p.), maximum flocculation of the clay suspension occurs, and this is accompanied by a maximum in σ_β. On further increase in CTAB concentration, further adsorption of CTA^+ occurs resulting in charge reversal and restabilization of the clay suspension. This is accompanied by reduction in σ_β.

12.4.10.2 Flocculation of Sterically Stabilized Dispersions

Neville and Hunter [15] studied the flocculation of polymethylmethacrylate (PMMA) latex stabilized with poly(ethylene oxide) (PEO). Flocculation was induced

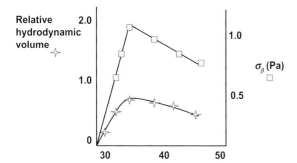

Figure 12.19 Variation of σ_β and hydrodynamic volume with temperature.

by addition of electrolyte and/or increase of temperature. Figure 12.19 shows the variation of σ_β with the increase of temperature at constant electrolyte concentration.

It can be seen that σ_β increases with the increase of temperature, reaching a maximum at the critical flocculation temperature (CFT) and then decreases with further increase in temperature. The initial increase is due to the flocculation of the latex with the increase of temperature, as result of reduction of solvency of the PEO chains with the increase of temperature. The reduction in σ_β after the CFT is due to the reduction in the hydrodynamic volume of the dispersion.

References

1 Ferry, J.D. (1980) *Viscoelastic Properties of Polymers*, John Wiley & Sons, Inc., New York.
2 Mackosko, C.W. (1994) *Rheology, Principles, Measurement and Applications*, John Wiley & Sons, Inc., New York.
3 Goodwin, J.W. (1984) *Surfactants*, (ed. T.F. Tadros) Academic Press, London.
4 Goodwin, J.W., and Hughes, R.W. (1992) *Adv. Colloid Interface Sci.*, **42**, 303.
5 Tadros, T.F. (1996) *Adv. Colloid Interface Sci.*, **68**, 97.
6 Goodwin, J.W., and Hughes, R.W. (2000) *Rheology for Chemists*, Royal Society of Chemistry Publication, Cambridge.
7 Wohrlow, R.W. (1980) *Rheological Techniques*, John Wiley & Sons, Inc., New York.
8 Bingham, E.C. (1922) *Fluidity and Plasticity*, McGraw-Hill, New York.
9 Herschel, W.H., and Bulkley, R. (1926) *Proc. Am. Soc. Test Mater.*, **26**, 621; *Kolloid Z.*, **39**, 291.
10 Casson, N. (1959) *Rheology of Disperse Systems* (ed. C.C. Mill), Pergamon Press, New York, pp. 84–104.
11 Cross, M.M. (1965) *J. Colloid Interface Sci.*, **20**, 417.
12 Tadros, T. (2010) *Rheology of Dispersions*, Wiley-VCH Verlag GmbH, Weinheim, Germany.
13 Buscall, R., Goodwin, J.W., Ottewill, R.H., and Tadros, T.F. (1982) *J. Colloid Interface Sci.*, **85**, 78.
14 Hunter, R.J., and Nicol, S.K. (1968) *J. Colloid Interface Sci.*, **28**, 250.
15 Neville, P.C., and Hunter, R.J. (1974) *J. Colloid Interface Sci.*, **49**, 204.

13
Rheology of Concentrated Suspensions

13.1
Introduction

Rheological measurements are useful tools for probing the microstructure of suspensions. This is particularly the case if measurements are carried out at low stresses or strains as discussed in Chapter 12. In this case, the special arrangement of particles is only slightly perturbed by the measurement. In other words, the convective motion due to the applied deformation is less than the Brownian diffusion. The ratio of the stress applied σ to the "thermal stress" (that is equal to $kT/6\pi a^3$, where k is the Boltzmann constant, T is the absolute temperature, and a is the particle radius) is defined in terms of a dimensionless Peclet number Pe:

$$Pe = \frac{6\pi a^3 \sigma}{kT} \tag{13.1}$$

For a colloidal particle with radius of 100 nm, σ should be less than 0.2 Pa to ensure that the microstructure is relatively undisturbed. In this case, $Pe < 1$.

In order to remain in the linear viscoelastic region, the structural relaxation by diffusion must occur on a time scale comparable to the experimental time. As mentioned in Chapter 4, the ratio of the structural relaxation time to the experimental measurement time is given by the dimensionless Deborah number De, which is ~1 and the suspension appears viscoelastic.

The rheology of suspensions depends on the balance between three main forces: Brownian diffusion; hydrodynamic interaction; and interparticle forces. These forces are determined by three main parameters: (i) the volume fraction ϕ (total volume of the particles divided by the volume of the dispersion); (ii) the particle size and shape distribution; and (iii) The net energy of interaction G_T, that is, the balance between repulsive and attractive forces.

The earliest theory for prediction of the relationship between the relative viscosity η_r and ϕ was described by Einstein, which is applicable to $\phi \leq 0.01$.

Dispersion of Powders in Liquids and Stabilization of Suspensions, First Edition. Tharwat F. Tadros.
© 2012 Wiley-VCH Verlag GmbH & Co. KGaA. Published 2012 by Wiley-VCH Verlag GmbH & Co. KGaA.

13.1.1
The Einstein Equation

Einstein [1] assumed that the particles behave as hard spheres (with no net interaction). The flow field has to dilate because the liquid has to move around the flowing particles. At $\phi \leq 0.01$, the disturbance around one particle does not interact with the disturbance around another. η_r is related to ϕ by the following expression [1]:

$$\eta_r = 1 + [\eta]\phi = 1 + 2.5\phi \tag{13.2}$$

where $[\eta]$ is referred to as the intrinsic viscosity and has the value 2.5.

For the above hard sphere very dilute dispersions, the flow is Newtonian, that is, the viscosity is independent of shear rate. At higher ϕ ($0.2 > \phi > 0.1$) values one has to consider the hydrodynamic interaction suggested by Batchelor [2] that is still valid for hard spheres.

13.1.2
The Batchelor Equation [2]

When $\phi > 0.01$, hydrodynamic interaction between the particles become important. When the particles come close to each other, the nearby stream lines and the disturbance of the fluid around one particle interact with that around a moving particle.

Using the above picture, Batchelor [2] derived the following expression for the relative viscosity:

$$\eta_r = 1 + 2.5\phi + 6.2\phi^2 + O\phi^3 \tag{13.3}$$

The third term in Eq. (13.3), that is, $6.2\phi^2$, is the hydrodynamic term whereas the fourth term is due to higher order interactions.

13.1.3
Rheology of Concentrated Suspensions

When $\phi > 0.2$, η_r becomes a complex function of ϕ. At such high-volume fractions, the system mostly shows non-Newtonian flow ranging from viscous to viscoelastic to elastic response. Three responses can be considered: (i) viscous response; (ii) elastic response; and (iii) viscoelastic response. These responses for any suspension depend on the time or frequency of the applied stress or strain (see Chapter 12).

Four different types of systems (with increasing complexity) can be considered as described below.

1) **Hard-sphere suspensions:** these are systems where both repulsive and attractive forces are screened.

2) **Systems with "soft" interaction:** these are systems containing electrical double layers with long-range repulsion. The rheology of the suspension is determined mainly by the double-layer repulsion.

3) **Sterically stabilized suspensions:** the rheology is determined by the steric repulsion produced by adsorbed nonionic surfactant or polymer layers – the interaction can be "hard" or "soft" depending on the ratio of adsorbed layer thickness to particle radius (δ/R).

4) **Flocculated systems:** these are systems where the net interaction is attractive.

One can distinguish between weak (reversible) and strong (irreversible) flocculation depending on the magnitude of the attraction.

13.1.3.1 Rheology of Hard-Sphere Suspensions

Hard-sphere suspensions (neutral stability) were developed by Krieger and coworkers [3, 4] using polystyrene latex suspensions whereby the double-layer repulsion was screened by using NaCl or KCl at a concentration of $10^{-3}\,mol\,dm^{-3}$ or replacing water by a less-polar medium such as benzyl alcohol.

The relative viscosity η_r ($= \eta/\eta_0$) is plotted as a function of reduced shear rate (shear rate × time for a Brownian diffusion t_r):

$$\dot{\gamma}_{red} = \dot{\gamma} t_r = \frac{6\pi \dot{\gamma} a^3}{kT} \tag{13.4}$$

where a is the particle radius, η_0 is the viscosity of the medium, k is the Boltzmann constant, and T is the absolute temperature.

A plot of (η/η_0) versus ($\eta_0 a^3/kT$) is shown in Figure 13.1 at $\phi = 0.4$ for particles with different sizes. At a constant ϕ, all points fall on the same curve. The curves are shifted to higher values for larger ϕ and to lower values for smaller ϕ.

The curve in Figure 13.1 shows two limiting (Newtonian) viscosities at low and high shear rates that are separated by a shear thinning region. In the low shear rate regime, the Brownian diffusion predominates over hydrodynamic interaction, and the system shows a "disordered" three-dimensional structure with high relative viscosity. As the shear rate is increased, this disordered structure starts to form layers coincident with the plane of shear, and this results in the shear thinning

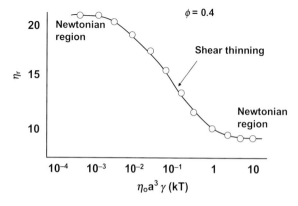

Figure 13.1 Reduced viscosity versus reduced shear rate for hard-sphere suspensions.

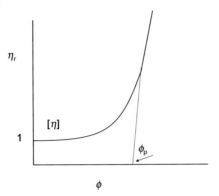

Figure 13.2 Relative viscosity versus volume fraction for hard-sphere suspensions.

region. In the high shear rate regime, the layers can "slide" freely, and hence a Newtonian region (with much lower viscosity) is obtained. In this region, the hydrodynamic interaction predominates over the Brownian diffusion.

If the relative viscosity in the first or second Newtonian region is plotted versus the volume fraction, one obtains the curve shown in Figure 13.2.

The curve in Figure 13.2 has two asymptotes: the slope of the linear portion at low ϕ values (the Einstein region) that gives the intrinsic viscosity $[\eta]$ that is equal to 2.5. The asymptote that occurs at a critical volume fraction ϕ_p at which the viscosity shows a sharp increase with increase in ϕ. ϕ_p is referred to as the maximum packing fraction for hard spheres: for hexagonal packing of equal-sized spheres $\phi_p = 0.74$. For random packing of equal-sized spheres, $\phi_p = 0.64$. For polydisperse systems, ϕ_p reaches higher values as is illustrated in Figure 13.3 for one-size, two-sizes, three-sizes, and four-sizes suspensions.

Analysis of the Viscosity–Volume Fraction Curve The best analysis of the η_r–ϕ curve is due to Dougherty and Krieger [3] who used a mean-field approximation by calculating the increase in viscosity as small increments of the suspension are consecutively added. Each added increment corresponds to replacement of the medium by more particles.

They arrived at the following simple semiempirical equation that could fit the viscosity data over the whole volume fraction range:

$$\eta_r = \left(1 - \frac{\phi}{\phi_p}\right)^{-[\eta]\phi_p} \tag{13.5}$$

Equation (13.4) is referred to as the Dougherty–Krieger Eq. (13.3) and is commonly used for analysis of the viscosity data.

13.1.3.2 Rheology of Systems with "Soft" or Electrostatic Interaction

In this case, the rheology is determined by the double-layer repulsion particularly with small particles and extended double layers [5]. In the low shear rate regime,

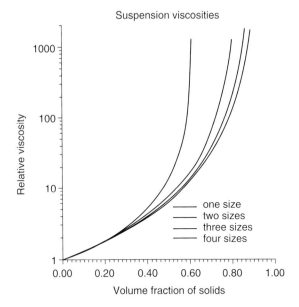

Figure 13.3 Viscosity–volume fraction curves for polydisperse suspensions.

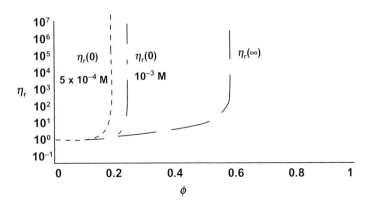

Figure 13.4 Variation of η_r with ϕ for polystyrene latex dispersions at two NaCl concentrations.

the viscosity is determined by the Brownian diffusion, and the particles approach each other to a distance of the order of ~4.5 κ^{-1} (where κ^{-1} is the "double-layer thickness" that is determined by electrolyte concentration and valency). This means that the effective radius of the particles R_{eff} is much higher than the core radius R. For example, for 100 nm particles with a zeta potential ζ of 50 mV dispersed in a medium of 10^{-5} mol dm^{-3} NaCl (κ^{-1} = 100 nm), R_{eff} ~ 325 nm. The effective volume fraction ϕ_{eff} is also much higher than the core volume fraction. This results in rapid increase in the viscosity at low-core volume fraction [5]. This is illustrated in Figure 13.4 which shows the variation of η_r with ϕ at 5×10^{-4} and

10^{-3} mol dm^{-3} NaCl ($R = 85$ nm and $\zeta = 78$ mV). The low shear viscosity $\eta_r(0)$ shows a rapid increase at $\phi \sim 0.2$ (the increase occurs at higher volume fraction at the higher electrolyte concentration). At $\phi > 0.2$, the system shows "solid-like" behavior with $\eta_r(0)$ reaching very high values (>10^7). At such high ϕ values, the system shows near plastic flow.

In the high shear rate regime, the increase in η_r occurs at much higher ϕ values. This is illustrated from the plot of the high shear relative viscosity $\eta_r(\infty)$ versus ϕ. At such high shear rates, the hydrodynamic interaction predominates over the Brownian diffusion, and the system shows a low viscosity denoted by $\eta_r(\infty)$. However when ϕ reaches a critical value, pseudoplastic flow is observed.

13.1.3.3 Rheology of Sterically Stabilized Dispersions

These are dispersions where the particle repulsion results from the interaction between adsorbed or grafted layers of nonionic surfactants or polymers [6–10]. The flow is determined by the balance of viscous and steric forces. The steric interaction is repulsive as long as the Floury–Huggins interaction parameter $\chi < \frac{1}{2}$. With short chains, the interaction may be represented by a hard-sphere type with $R_{\text{eff}} = R + \delta$. This is particularly the case with nonaqueous dispersions with an adsorbed layer of thickness smaller compared to the particle radius (any electrosatic repulsion is negligible in this case). With most sterically stabilized dispersions, the adsorbed or grafted layer has an appreciable thickness (compared to particle radius), and hence the interaction is "soft" in nature as a result of the longer range of interaction. Results for aqueous sterically stabilized dispersions were produced using polystyrene (PS) latex with grafted poly(ethylene oxide) (PEO) layers [11–17]. As an illustration, Figure 13.5 shows the variation of η_r with ϕ for latex dispersions with three particle radii (77.5, 306, and 502 nm). For comparison, the η_r–ϕ curve calculated using the Dougherty–Krieger equation is shown in the same figure. The η_r–ϕ curves are shifted to the left as a result of the presence of the grafted PEO layers. The experimental relative viscosity data may be used to

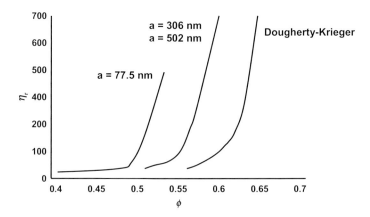

Figure 13.5 η_r–ϕ curves for PS latex dispersions containing grafted PEO chains.

obtain the grafted polymer layer thickness at various volume fractions of the dispersions. Using the Dougherty–Krieger equation, one can obtain the effective volume fraction of the dispersion. From the knowledge of the core volume fraction, one can calculate the grafted layer thickness at each dispersion volume fraction.

To apply the Dougherty–Krieger equation, one needs to know the maximum packing fraction, ϕ_p. This can be obtained from a plot of $1/(\eta_r)^{1/2}$ versus ϕ and extrapolation to $1/(\eta_r)^{1/2}$, using the following empirical equation:

$$\frac{K}{\eta_r^{1/2}} = \phi_p - \phi \tag{13.6}$$

The value of ϕ_p using Eq. (13.12) was found to be in the range 0.6–0.64. The intrinsic viscosity $[\eta]$ was assigned a value of 2.5.

Using the above calculations, the grafted PEO layer thickness δ was calculated as a function of ϕ for the three latex dispersions. For the dispersions with $R = 77.5$ nm, δ was found to be 8.1 nm at $\phi = 0.42$, decreasing to 5.0 nm when ϕ was increased to 0.543. For the dispersions with $R = 306$ nm, $\delta = 12.0$ nm at $\phi = 0.51$, decreasing to 10.1 when ϕ was increased to 0.60. For the dispersions with $R = 502$ nm, $\delta = 21.0$ nm at $\phi = 0.54$ decreases to 14.7 as ϕ was increased to 0.61.

Viscoelastic Properties of Sterically Stabilized Suspensions The rheology of sterically stabilized dispersions is determined by the steric repulsion particularly for small particles with "thick" adsorbed layers. This is illustrated in Figures 13.6 and 13.7, which shows the variation of G^*, G', and G'' with frequency (Hz) for poystyrene latex dispersions of 175 nm radius containing grafted poly(ethylene oxide) (PEO) with a molecular weight of 2000 (giving a hydrodynamic thickness $\delta \sim 20$ nm) [12]. The results clearly show the transition from predominantly viscous response when $\phi \leq 0.465$ to predominantly elastic response when $\phi \geq 0.5$. This behavior reflects the steric interaction between the PEO layers. When the surface-to-surface distance between the particles h becomes $<2\delta$, the elastic interaction occurs and $G' > G''$.

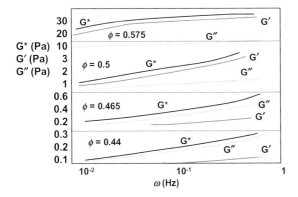

Figure 13.6 Variation of G^*, G', and G'' with frequency for sterically stabilized dispersions.

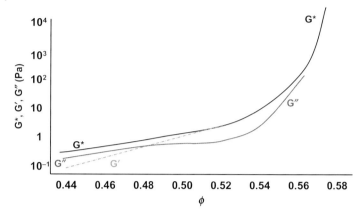

Figure 13.7 Variation of G^*, G', and G'' (at $\omega = 1\,\text{Hz}$) with ϕ for latex dispersions ($a = 175\,\text{nm}$) containing grafted PEO chains.

The exact volume fraction at which a dispersion changes from predominantly viscous to predominantly elastic response may be obtained from plots of G^*, G', and G'' (at fixed strain in the linear viscoelastic region and fixed frequency) versus the volume fraction of the dispersion. This is illustrated in Figure 13.7, which shows the results for the above latex dispersions. At $\phi = 0.482$, $G' = G''$ (sometimes referred to as the cross-over point), which corresponds to $\phi_{\text{eff}} \sim 0.62$ (close to maximum random packing). At $\phi > 0.482$, G' becomes progressively larger than G'', and ultimately the value of G' approaches G^* and G'' becomes relatively much smaller than G'. At $\phi = 0.585$, $G' \sim G^* = 4.8 \times 10^3$ and at $\phi = 0.62$, $G' \sim G^* = 1.6 \times 10^5\,\text{Pa}$. Such high elastic moduli values indicate that the dispersions behave as near elastic solids ("gels") as a result of interpenetration and/or compression of the grafted PEO chains.

13.1.3.4 Rheology of Flocculated Suspensions

The rheology of unstable systems poses problems both from the experimental and theoretical points of view. This is due to the nonequilibrium nature of the structure, resulting from the weak Brownian motion [18]. For this reason, advances on the rheology of suspensions, where the net energy is attractive, have been slow and only of qualitative nature. On the practical side, control of the rheology of flocculated and coagulated suspensions is difficult since the rheology depends not only on the magnitude of the attractive energies but also on how one arrives at the flocculated or coagulated structures in question. As mentioned in Chapter 9, various structures can be formed, for example, compact flocs, weak and metastable structures, chain aggregates, etc. At high volume fraction of the suspension, a "three-dimensional" network of the particles is formed throughout the sample. Under shear this network is broken into smaller units of flocculated spheres which can withstand the shear forces [19]. The size of the units that survive is determined by the balance of shear forces which tend to break the units down and the energy of attraction that holds the spheres together [20–22]. The appropriate dimensionless group characterizing this process (balance of viscous and van der Waals force) is $\eta_0 a^4 \dot{\gamma}/A$ (where η_0 is the viscosity of the medium, a is the particle radius, $\dot{\gamma}$ is the shear rate, and A is the effective Hamaker constant). Each flocculated unit is expected to rotate in the shear field, and it is likely that these units will tend to form layers as individual spheres do. As the shear stress increases, each rotating unit will ultimately behave as an individual sphere and, therefore, a flocculated suspension will show pseudoplastic flow with the relative viscosity approaching a constant value at high shear rates. The viscosity-shear rate curve will also show a pseudo-Newtonian region at low and high shear rates (similar to the case with stable systems described above). However, the values of the low and high shear rate viscosities (η_0 and η_∞) will of course depend on the extent of flocculation and the volume fraction of the suspension. It is also clear that such systems will show an apparent yield stress (Bingham yield value, σ_β) normally obtained by extrapolation of the linear portion of the $\sigma - \dot{\gamma}$ curve to $\dot{\gamma} = 0$. Moreover, since the structural units of a weakly flocculated system change with the change in the shear rate, most flocculated suspensions show thixotropy as discussed in Chapter 12. Once shear is initiated, some finite time is required to break the network of agglomerated flocs into smaller units which persist under the shear forces applied. As smaller units are formed, some of the liquid entrapped in the flocs is liberated, thereby reducing the effective volume fraction, ϕ_{eff}, of the suspension. This reduction in ϕ_{eff} is accompanied by a reduction in η_{eff}, and this plays a major role in generating thixotropy.

It is convenient to distinguish between two type of unstable systems depending on the magnitude of the net attractive energy. (i) Weakly flocculated suspensions: the attraction in this case is weak (energy of few kT units) and reversible, for example, in the secondary minimum of the DLVO curve or the shallow minimum obtained with sterically stabilized systems. A particular case of weak flocculation is that obtained on the addition of "free" (nonadsorbing) polymer referred to as depletion flocculation. (ii) Strongly flocculated (coagulated) suspensions: the attraction in this case is strong (involving energies of several $100 kT$ units) and irreversible. This is the case of flocculation in the primary minimum or those flocculated by reduction of solvency of the medium (for sterically stabilized suspensions) to worse than a θ-solvent. The study of the rheology of flocculated suspensions is difficult since the structure of the flocs is at nonequilibrium. Theories for flocculated suspensions are also qualitative and based on a number of assumptions.

Weakly Flocculated Suspensions As mentioned in Chapter 9, weak flocculation may be obtained by the addition of "free" (nonadsorbing) polymer to a sterically stabilized dispersion [10]. Several rheological investigations of such systems have been carried out by Tadros and his collaborators [23–27]. This is exemplified by a latex dispersion containing grafted PEO chains of $M = 2000$ to which "free" PEO is added at various concentrations. The grafted PEO chains that were made sufficiently dense ensure the absence of adsorption of the added free polymer. Three

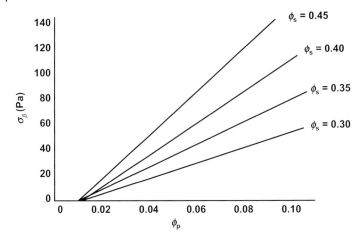

Figure 13.8 Variation of yield value σ_β with volume fraction ϕ_p of "free polymer" (PEO; M = 20 000) at various latex volume fractions ϕ_s.

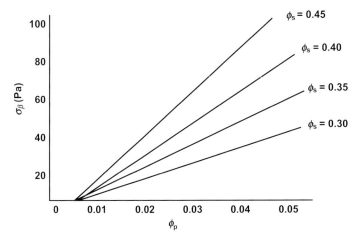

Figure 13.9 Variation of yield value σ_β with volume fraction ϕ_p of "free polymer" (PEO; M = 35 000) at various latex volume fractions ϕ_s.

molecular weight PEOs were used: 20 000, 35 000, and 90 000. As an illustration, Figures 13.8–13.10 show the variation of the Bingham yield value σ_β with volume fraction of "free" polymer ϕ_p at the three PEO molecular weights studied and at various latex volume fractions ϕ_s. The latex radius R in this case was 73.5 nm. The results of Figures 13.8–13.10 show a rapid and linear increase in σ_β with the increase in ϕ_p when the latter exceeds a critical value, ϕ_p^+. The latter is the critical free polymer volume fraction for depletion flocculation. ϕ_p^+ decreases with the increase of the molecular weight M of the free polymer, as expected. There does not seem to be any dependence of ϕ_p^+ on the volume fraction of the latex, ϕ_s.

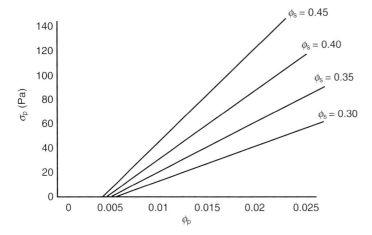

Figure 13.10 Variation of yield value σ_β with volume fraction ϕ_p of "free polymer" (PEO; M = 100 000) at various latex volume fractions ϕ_s.

Table 13.1 Volume fraction of free polymer at which flocculation starts, ϕ_p^+.

Particle radius (nm)	M (PEO)	ϕ_p^+
73.5	20 000	0.0150
73.5	35 000	0.0060
73.5	10 000	0.0055
73.5	20 000	0.0150
217.5	20 000	0.0055
457.5	20 000	0.0050

Similar trends were obtained using larger latex particles (with radii 217.5 and 457.5 nm). However, there was a definite trend of the effect of particle size; the larger the particle size, the smaller the value of ϕ_p^+. A summary of ϕ_p^+ for the various molecular weights and particle sizes is given in Table 13.1.

The results in Table 13.1 show a significant reduction in ϕ_p^+ when the molecular weight of PEO is increased from 20 000 to 35 000, whereas when M is increased from 35 000 to 100 000, the reduction in ϕ_p^+ is relatively smaller. Similarly, there is a significant reduction in ϕ_p^+ when the particle radius is increased from 73.5 to 217.5 nm, with a relatively smaller decrease on further increase of a to 457.5 nm.

The straight line relationship between the extrapolated yield value and the volume fraction of free polymer can be described by the following scaling law:

$$\sigma_\beta = K\varphi_s^m \left(\varphi_p - \varphi_p^+ \right) \qquad (13.7)$$

Table 13.2 Power law plot for σ_β versus ϕ_s for various PEO molecular weights and latex radii.

Latex $R = 73.5$ nm

PEO 20 000		PEO 35 000		PEO 100 000	
ϕ_p	m	ϕ_p	m	ϕ_p	m
0.040	3.0	0.022	2.9	0.015	2.7
0.060	2.7	0.030	3.0	0.020	2.7
0.080	2.8	0.040	2.8	0.025	2.8
0.100	2.8	0.050	2.9	–	–

Latex $R = 217.5$ | | **Latex $R = 457.5$** | |

ϕ_p	m	ϕ_p	m
0.020	3.0	0.020	2.7
0.040	2.9	0.030	2.7
0.060	2.8	0.040	2.8
0.080	2.8	0.050	2.7

where K is a constant and m is the power exponent in ϕ_s, which may be related to the flocculation process. The values of m used to fit the data of σ_β versus ϕ_s are summarized in Table 13.2.

It can be seen from Table 13.2 that m is nearly constant, being independent of particle size and free polymer concentration. Age value for m of 2.8 may be assigned for such weakly flocculated system. This value is close to the exponent predicted for diffusion-controlled aggregation (3.5 ± 0.2) predicted by Ball and Brown [28, 29] who developed a computer simulation method treating the flocs as fractals that are closely packed throughout the sample.

The near independence of ϕ_p^+ on ϕ_s can be explained on the basis of the statistical mechanical approach of Gast et al. [30], which showed such independence when the osmotic pressure of the free polymer solution is relatively low and/or the ratio of the particle diameter to the polymer coil diameter is relatively large (>8–9). The latter situation is certainly the case with the latex suspensions with diameters of 435 and 915 nm at all PEO molecular weights. The only situation where this condition is not satisfied is with the smallest latex and the highest molecular weight.

The dependence of ϕ_p^+ on particle size can be explained from a consideration of the dependence of free energy of depletion and van der Waals attraction on particle radius as will be discussed below. Both attractions increase with the increase of particle radius. Thus the larger particles would require smaller free polymer concentration at the onset of flocculation.

It is possible, in principle, to relate the extrapolated Bingham yield value, σ_β, to the energy required to separate the flocs into single units, E_{sep} [23, 24]:

Table 13.3 Results of E_{sep}, G_{dep} calculated on the basis of AO and FSV models.

ϕ_p	ϕ_s	σ_β (Pa)	E_{sep}/kT	G_{dep}/kT	
				AO model	FSV model
(a) M (PEO) = 20 000					
0.04	0.30	12.5	8.4	18.2	78.4
	0.35	21.0	12.1	18.2	78.4
	0.40	30.5	15.4	18.2	78.4
	0.45	40.0	18.0	18.2	78.4
(b) M (PEO) = 35 000					
0.03	0.30	17.5	11.8	15.7	78.6
	0.35	25.7	14.8	15.7	78.6
	0.40	37.3	18.9	15.7	78.6
	0.45	56.8	25.5	15.7	78.6
(c) M (PEO) = 10 000					
0.02	0.30	10.0	6.7	9.4	70.8
	0.35	15.0	8.7	9.4	70.8
	0.40	22.0	11.1	9.4	70.8
	0.45	32.5	14.6	9.4	70.8

$$\sigma_\beta = \frac{3\phi_s n E_{sep}}{8\pi R^3} \tag{13.8}$$

where n is the average number of contacts per particle (the coordination number). The maximum value of n is 12 which corresponds to hexagonal packing of the particles in a floc. For random packing of particles in the floc, $n = 8$. However, it is highly unlikely that such values of 12 or 8 are reached in a weakly flocculated system and a more realistic value for n is probably 4 (a relatively open structure in the floc).

In order to calculate E_{sep} from σ_β, one has to assume that all particle–particle contacts are broken by shear. This is highly likely since the high shear viscosity of the weakly flocculated latex was close to that of the latex before the addition of the free polymer. The values of E_{sep} obtained using Eq. (13.19) with $n = 4$ are given in Table 13.3 at the three PEO molecular weights for the latex with the radius of 73.5 nm. It can be seen that E_{sep}, at any given ϕ_p, increases with the increase of the volume fraction ϕ_s of the latex.

A comparison between E_{sep} and the free energy of depletion flocculation, G_{dep}, can be made using the theories of Asakura and Oosawa (AO) [31, 32] and Fleer, Vincent, and Scheutjenss (FVS) [33]. Asakura and Oosawa [32, 33] derived the following expression for G_{dep}, which is valid for the case where the particle radius is much larger than the polymer coil radius:

$$\frac{G_{dep}}{kT} = -\frac{3}{2}\varphi_2\beta x^2; \quad 0 < x < 1 \tag{13.9}$$

where k is the Boltzmann constant, T is the absolute temperature, ϕ_2 is the volume concentration of free polymer that is given by

$$\phi_2 = \frac{4\pi\Delta^3 N_2}{3v} \tag{13.10}$$

Δ is the depletion layer thickness that is equal to the radius of gyration of free polymer, and R_g and N_2 is the total number of polymer molecules in a volume v of solution:

$$\beta = \frac{R}{\Delta} \tag{13.11}$$

$$x = \frac{[\Delta - (h/2)]}{\Delta} \tag{13.12}$$

where h is the distance of separation between the outer surfaces of the particles. Clearly when $h = 0$, that is, at the point where the polymer coils are "squeezed out" from the region between the particles, $x = 1$.

Fleer, Scheutjens, and Vincent (FSV model) [32] developed a general approach of the interaction of hard spheres in the presence of a free polymer. This model takes into account the dependence of the range of interaction on free polymer concentration and any contribution from the nonideal mixing of polymer solutions. This theory gives the following expression for G_{dep}:

$$G_{dep} = 2\pi R \left(\frac{\mu_1 - \mu_1^0}{v_1^0} \right) \Delta^2 \left(1 + \frac{2\Delta}{3R} \right) \tag{13.13}$$

where μ_1 is the chemical potential at bulk polymer concentration ϕ_p, μ_1^0 is the corresponding value in the absence of free polymer, and v_1^0 is the molecular volume of the solvent.

The difference in chemical potential $(\mu_1 - \mu_1^0)$ can be calculated from the volume fraction of free polymer ϕ_p and the polymer–solvent (Flory–Huggins) interaction parameter χ ():

$$\frac{\mu_1 - \mu_1^0}{kT} = -\left[\frac{\varphi_p}{n_2} + \left(\frac{1}{2} - \chi \right) \varphi_p^2 \right] \tag{13.14}$$

where n_2 is the number of polymer segments per chain.

A summary of the values of E_{sep}, G_{dep}, calculated on the basis of AO and FSV models is given in Table 13.3 at three molecular weights for PEO and for a latex with $a = 77.5$ nm.

It can be seen from Table 13.3 that E_{sep}, at any given ϕ_p, increases with the increase of the volume fraction ϕ_s of the latex. In contrast, the value of G_{dep} does not depend on the value of ϕ_s. The theories on depletion flocculation only show a dependence of G_{dep} on ϕ_p and a. Thus, one cannot make a direct comparison between E_{sep} and G_{dep}. The close agreement between E_{sep} and G_{dep} using Asakura and Oosawa's theory [31, 32] and assuming a value of $n = 4$ should only be considered fortuitous.

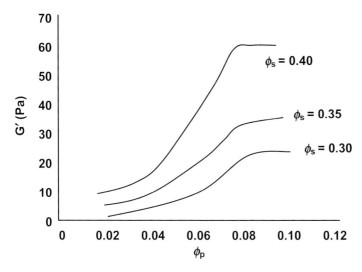

Figure 13.11 Variation of storage Mondulus' G' with volume fraction of polymer (PEO; M = 20 000).

Using Eqs. (13.18) and (13.19), a general scaling law may be used to show the variation of E_{sep} with the various parameters of the system:

$$E_{sep} = \frac{8\pi R^3}{3\varphi_s n} K_1 \varphi_s^{2.8} \left(\varphi_p - \varphi_p^+\right) = \frac{8\pi K_1}{3n} R^3 \varphi_s^{1.8} \left(\varphi_p - \varphi_p^+\right) \quad (13.15)$$

Equation (13.15) shows the four parameters that determine E_{sep}: the particle radius a, the volume fraction of the suspension ϕ_s, the concentration of free polymer ϕ_p, and the molecular weight of the free polymer, which together with a determines ϕ_p^+.

More insight on the structure of the flocculated latex dispersions was obtained using viscoelastic measurements [26]. As an illustration, Figure 13.11 shows the variation of the storage modulus G' with ϕ_p ($M = 20\,000$) at various latex ($a = 77.5$) volume fractions ϕ_s. Similar trends were obtained for the other PEO molecular weight. All results show the same trend, namely an increase in G' with the increase in ϕ_p reaching a plateau value at high ϕ_p values. These results are different from those obtained using steady-state measurements, which show a rapid and linear increase of yield value σ_β. This difference reflects the behavior when using oscillatory (low-deformation) measurements which causes little perturbation of the structure when using low-amplitude and high-frequency measurements. Above ϕ_p^+ flocculation occurs and G' increases in magnitude with further increase in ϕ_p until a three-dimensional network structure is reached and G' reaches a limiting value. Any further increase in free polymer concentration may cause a change in the floc structure but this may not cause a significant increase in the number of bonds between the units formed (which determine the magnitude of G').

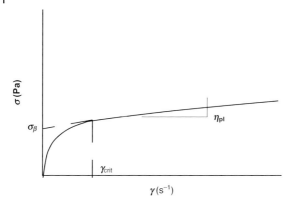

Figure 13.12 Pseudoplastic flow curve for a flocculated suspension.

Strongly Flocculated (Coagulated) Suspensions Steady-state shear stress–shear rate curves show pseudoplastic flow curve as is illustrated in Figure 13.12. The flow curve is characterized by three main parameters: (i) the shear rate above which the flow curve shows linear behavior. Above this shear rate collisions occur between the flocs, and this may cause interchange between the flocculi (the smaller floc units that aggregate to form a floc). In this linear region, the ratio of the floc volume to the particle volume (ϕ_F/ϕ_p) that is, the floc density remains constant. (ii) Next is σ_β the residual stress (yield stress) that arises from the residual effect of inter-particle potential. (iii) Final is η_{pl}, the slope of the linear portion of the flow curve that arises from purely hydrodynamic effects.

13.1.4
Analysis of the Flow Curve

13.1.4.1 Impulse Theory: Goodeve and Gillespie [34, 35]

The interparticle interaction effects (given by σ_β) and hydrodynamic effects (give by η_{pl}) are assumed to be additive:

$$\sigma = \sigma_\beta + \eta_{pl}\dot{\gamma} \quad (13.16)$$

To calculate σ_β, Goodeve proposed that when shearing occurs, links between particles in a flocculated structure are stretched, broken, and reformed. An impulse is transferred from a fast-moving layer to a slow-moving layer. Non-Newtonian effects are due to the effect of shear on the number of links, the lifetime of a link, and any change in the size of the floc.

According to Goodeve theory, the yield value is given by

$$\sigma_\beta = \left(\frac{3\phi^2}{2\pi R^3}\right) E_A \quad (13.17)$$

13.1 Introduction

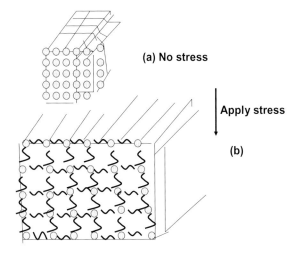

Figure 13.13 Schematic picture of the elastic floc.

where ϕ is the volume fraction of the dispersed phase, a is the particle radius, and E_A is the total binding energy.

$$E_A = n_L \varepsilon_L \tag{13.18}$$

where n_L is the number of links with a binding energy ε_L per link.

According to Eq. (13.6): $\sigma\beta \propto \phi^2$; $\propto (1/a^3)$; $\propto E_A$ (the energy of attraction).

13.1.4.2 Elastic Floc Model: Hunter and Coworkers [36, 37]

The floc is assumed to consist of an open network of "girders" as schematically shown in Figure 13.13. The floc undergoes extension and compression during rotation in a shear flow. The bonds are stretched by a small amount Δ (that can be as small as 1% of particle radius).

To calculate σ_β, Hunter considered the energy dissipation during rupture of the flocs (assumed to be consist of doublets).

The yield value σ_β is given by the expression

$$\sigma_\beta = \alpha_0 \beta \lambda \eta \dot{\gamma} \left(\frac{R_{\text{floc}}^2}{R^3} \right) \phi_s^2 \Delta C_{\text{FP}} \tag{13.19}$$

where α_0 is the collision frequency, β is a constant (= (27/5)), λ is a correction factor (~1), and CFP is the floc density (= ϕ_F/ϕ_s).

13.1.5 Fractal Concept for Flocculation

The floc structure can be treated as fractals whereby an isolated floc with radius a_F can be assumed to have uniform packing throughout that floc [38, 39].

In the above case, the number of particles in a floc is given by

$$n_F = \phi_{mf} \left(\frac{R_F}{R}\right)^3 \tag{13.20}$$

where ϕ_{mf} is the packing fraction of the floc.

If the floc does not have constant packing throughout its structure but is dentritic in form, the packing density of the floc begins to reduce as one goes from the center to the edge. If this reduction is with a constant power law D,

$$n_F = \left(\frac{R_F}{R}\right)^D \tag{13.21}$$

where $0 < D \leq 3$.

D is called the packing index, and it represents the packing change with distance from the center. Two cases may be considered.

i) **Rapid aggregation:** (diffusion limited aggregation (DLA)). When particles touch they stick particle–particle aggregation gives $D = 2.5$; aggregate–aggregate.

   Aggregation gives $D = 1.8$. The lower the value of D, the more open the floc structure is.

ii) **Slow aggregation:** (rate limited aggregation (RLA)). The particles have a lower sticking probability – some are able to rearrange and densify the floc – $D \approx 2.0$–2.2.

   The lower the value of D, the more open the floc structure is. Thus by determining D, one can obtain information on the flocculation behavior. If flocculation of a suspension occurs by changing the conditions (e.g., increasing temperature), one can visualize sites for nucleation of flocs occurring randomly throughout the whole volume of the suspension.

The total number of primary particles does not change, and the volume fraction of the floc is given by

$$\phi_F = \phi \left(\frac{R_F}{F}\right)^{3-D} \tag{13.22}$$

Since the yield stress σ_β and elastic modulus G' depend on the volume fraction, one can use a power law in the form:

$$\sigma_\beta = K\phi^m \tag{13.23}$$

$$G' = K\phi^m \tag{13.24}$$

where the exponent m reflects the fractal dimension.

Thus by plotting $\log \sigma_\beta$ or $\log G'$ versus $\log \phi$, one can obtain m from the slope which can be used to characterize the floc nature and structure, $m = 2/(3-D)$.

13.1.6
Examples of Strongly Flocculated (Coagulated) Suspension

13.1.6.1 Coagulation of Electrostatically Stabilized Suspensions by Addition of Electrolyte

As mentioned in Chapter 4, electrostically stabilized suspensions become coagulated when the electrolyte concentration is increased above the critical coagulation concentration (CCC). This is illustrated by using a latex dispersion (prepared using surfactant-free emulsion polymerization) to which 0.2 mol dm^{-3} NaCl is added (well above the CCC which is 0.1 mol dm^{-3} NaCl).

Figure 13.14 shows the strain sweep results for latex dispersions at various volume fractions ϕ and in the presence of 0.2 mol dm^{-3} NaCl.

It can be seen from Figure 13.14 that G^* and G' (which are very close to each other) remain independent of the applied strain (the linear viscoelastic region), but above a critical strain, γ_{cr}, G^* and G' show a rapid reduction with further increase in strain (the nonlinear region). In contrast G'' (which is much lower than G') remains constant showing an ill-defined maximum at intermediate strains. Above γ_{cr}, the flocculated structure becomes broken down with applied shear.

Figure 13.15 shows the variation of G' (measured at strains in the linear viscoelastic region) with frequency v (in Hz) at various latex volume fractions. As mentioned above, G' is almost equal to G^* since G'' is very low.

In all cases, $G' \gg G''$ and it shows little dependence on frequency. This behavior is typical of highly elastic (coagulated) structure, whereby a "continuous gel" network structure is produced at such high volume fractions (see Chapter 9).

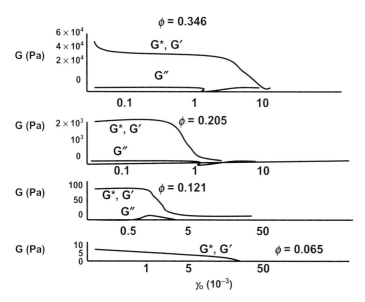

Figure 13.14 Strain sweep results for latex dispersions at various volume fractions in the presence of 0.2 mol dm^{-3} NaCl.

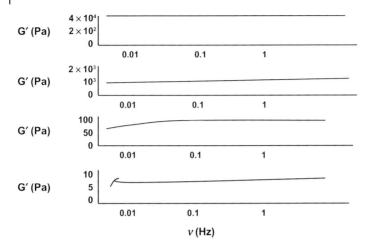

Figure 13.15 Variation of G with frequency at various latex volume fractions.

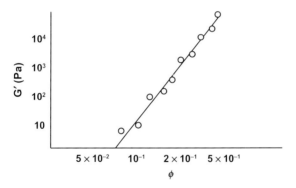

Figure 13.16 Log–log plots of G' versus ϕ for coagulated polystyrene latex suspensions.

Scaling laws can be applied for the variation of G' with the volume fraction of the latex ϕ. A log–log plot of G' versus ϕ is shown in Figure 13.16. This plot is linear and can be represented by the following scaling equation:

$$G' = 1.98 \times 10^7 \, \phi^{6.0} \tag{13.25}$$

The high power in ϕ is indicative of a relatively compact coagulated structure. This power gives a fractal dimension of 2.67 that confirms the compact structure.

It is also possible to obtain the cohesive energy of the flocculated structure E_c from a knowledge of G' (in the linear viscoelastic region) and γ_{cr}. E_c is related to the stress σ in the coagulated structure by the following equation:

$$E_c = \int_0^{\gamma_{cr}} \sigma \, d\gamma \tag{13.26}$$

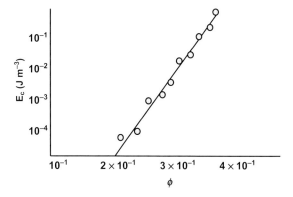

Figure 13.17 Log–log plots of E_c versus ϕ for coagulated polystyrene latex suspensions.

Since $\sigma = G'\gamma_0$, then

$$E_c = \int_0^{\gamma_{cr}} \gamma_0 G' \, d\gamma = \frac{1}{2}\gamma_{cr}^2 G' \qquad (13.27)$$

A log–log plot of E_c versus ϕ is shown in Figure 13.17 for such coagulated latex suspensions.

The E_c versus ϕ curve can be represented by the following scaling relationship:

$$E_c = 1.02 \times 10^3 \varphi^{9.1} \qquad (13.28)$$

The high power in ϕ is indicative of the compact structure of these coagulated suspensions.

13.1.7
Strongly Flocculated Sterically Stabilized Systems

13.1.7.1 Influence of the Addition of Electrolyte

As mentioned in Chapter 8, sterically stabilized suspensions show strong flocculation (sometimes referred to as incipient flocculation) when the medium for the stabilizing chain becomes worse than a θ-solvent (the Flory–Huggins interaction parameter, $\chi > 0.5$). Reduction of solvency for a PEO-stabilizing chain can be reduced by addition of a nonsolvent for the chains or by addition of electrolyte such as Na_2SO_4. Above a critical Na_2SO_4 concentration (to be referred to as the critical flocculation concentration, CFC), the χ parameter exceeds 0.5 and this results in incipient flocculation. This process of flocculation can be investigated using rheological measurements without diluting the latex. This dilution may result in change in the floc structure, and hence investigations without dilution ensure the absence of change of the floc structure, in particular, when using low-deformation (oscillatory) techniques [40].

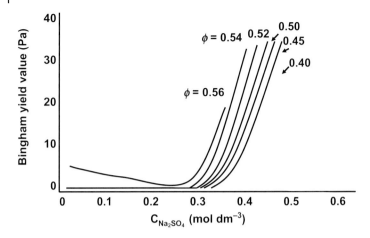

Figure 13.18 Variation of Bingham yield value with Na_2SO_4 concentration at various volume fractions ϕ of latex.

Figure 13.18 shows the variation of extrapolated yield value, σ_β, as a function of Na_2SO_4 concentration at various latex volume fractions ϕ_s at 25 °C. The latex had a z-average particle diameter of 435 nm, and it contained grafted PEO with $M = 2000$. It is clear that when $\phi_s < 0.52$, σ_β is virtually equal to zero up to $0.3\,mol\,dm^{-3}$ Na_2SO_4 above which it shows a rapid increase in σ_β with further increase in Na_2SO_4 concentration. When $\phi_s > 0.52$, a small yield value is obtained below $0.3\,mol\,dm^{-3}$ Na_2SO_4, which may be attributed to the possible elastic interaction between the grafted PEO chains when the particle–particle separation is less than 2δ (where δ is the grafted PEO layer thickness). Above $0.3\,mol\,dm^{-3}$ Na_2SO_4, there is a rapid increase in σ_β. Thus, the CFC of all concentrated latex dispersions is around $0.3\,mol\,dm^{-3}$ Na_2SO_4. It should be mentioned that at Na_2SO_4 below the CFC, σ_β shows a measurable decrease with increase of Na_2SO_4 concentration. This is due to the reduction in the effective radius of the latex particles as a result of the reduction in the solvency of the medium for the chains. This accounts for a reduction in the effective volume fraction of the dispersion which is accompanied by a reduction in σ_β.

Figure 13.19 shows the results for the variation of the storage modulus G' with Na_2SO_4 concentration. These results show the same trend as those shown in Figure 13.18, that is, an initial reduction in G' due to the reduction in the effective volume fraction, followed by a sharp increase above the CFC (which is $0.3\,mol\,dm^{-3}$ Na_2SO_4). Log–log plots of σ_β and G' versus ϕ_s at various Na_2SO_4 concentrations are shown in Figures 13.20 and 13.21. All the data are described by the following scaling equations:

$$\sigma_\beta = k\varphi_s^m \qquad (13.29)$$

$$G' = k'\varphi_s^n \qquad (13.30)$$

with $0.35 < \phi_s < 0.53$.

13.1 Introduction

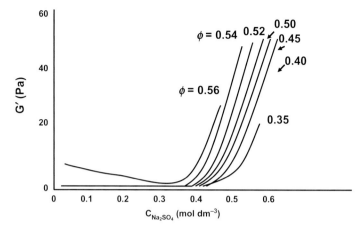

Figure 13.19 Variation of storage modulus G' with Na_2SO_4 concentration at various volume fractions ϕ of latex.

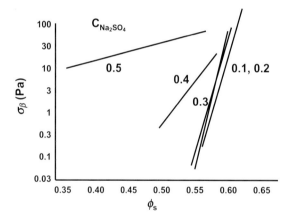

Figure 13.20 Log–log plots of σ_β versus ϕ_s.

The values of m and n are very high at Na_2SO_4 concentration below the CFC reaching values in the range of 30–50 which indicate the strong steric repulsion between the latex particles. When the Na_2SO_4 concentration exceeds the CFC, m and n decrease very sharply reaching a value of $m = 9.4$ and $n = 12$ when the Na_2SO_4 concentration increases to $0.4\,\mathrm{mol\,dm^{-3}}$ and a value of $m = 2.8$ and $n = 2.2$ at $0.5\,\mathrm{mol\,dm^{-3}}$ Na_2SO_4. These slopes can be used to calculate the fractal dimensions (see above) giving a value of $D = 2.70$–2.75 at $0.4\,\mathrm{mol\,dm^{-3}}$ and $D = 1.6$–1.9 at $0.5\,\mathrm{mol\,dm^{-3}}$ Na_2SO_4.

The above results of fractal dimensions indicate a different floc structure when compared with the results obtained using electrolyte to induce coagulation. In the latter case, $D = 2.67$ indicating a compact structure similar to that obtained at

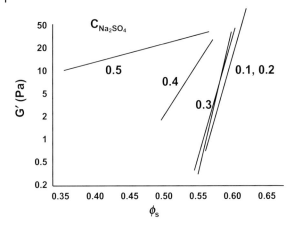

Figure 13.21 Log–log plots of G' versus ϕ_s.

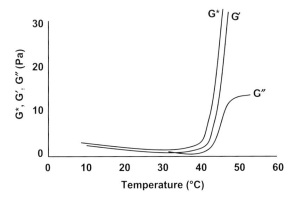

Figure 13.22 Variation of G^*, G', and G'' with temperature for latex dispersions ($\phi_s = 0.55$) in $0.2\,\mathrm{mol\,dm^{-3}}$ Na_2SO_4.

$0.4\,\mathrm{mol\,dm^{-3}}$ Na_2SO_4. However, when the Na_2SO_4 concentration exceeded the CFC value ($0.5\,\mathrm{mol\,dm^{-3}}$ Na_2SO_4), a much more open floc structure with a fractal dimension less than 2 was obtained.

13.1.7.2 Influence of Increase of Temperature

Sterically stabilized dispersions with PEO chains as stabilizers undergo flocculation on increasing the temperature. At a critical temperature (critical flocculation temperature, CFT), the Flory–Huggins interaction parameter becomes higher than 0.5 resulting in incipient flocculation. This is illustrated in Figure 13.22 which shows the variation of the storage modulus G' and loss modulus G'' with increasing temperature for a latex dispersion with a volume fraction $\phi = 0.55$ and at Na_2SO_4 concentration of $0.2\,\mathrm{mol\,dm^{-3}}$. At this electrolyte concentration, the latex is stable in the temperature range 10–40 °C. However, above this temperature (CFT), the latex is strongly flocculated.

The results of Figure 13.22 show an initial systematic reduction in the moduli values with increasing temperature up to 40 °C. This is the result of reduction in solvency of the chains with increasing temperature. The latter increase causes a breakdown in the hydrogen bonds between the PEO chains and water molecules. This results in a reduction in the thickness of the grafted PEO chains and hence a reduction in the effective volume fraction of the dispersion. The latter causes a decrease in the moduli values. However, at 40 °C, there is a rapid increase in the moduli values with further increase of temperature. The latter indicates the onset of flocculation (the CFT). Similar results were obtained at 0.3 and 0.4 mol dm^{-3} Na$_2$SO$_4$, but in these cases, the CFT was 35 and 15 °C, respectively.

13.1.8
Models for Interpretation of Rheological Results

13.1.8.1 Dublet Floc Structure Model

Neville and Hunter [41] introduced a doublet floc model to deal with sterically stabilized dispersions, which have undergone flocculation. They assumed that the major contribution to the excess energy dissipation in such pseudoplastic systems comes from the shear field which provides energy to separate contacting particles in a floc. The extrapolated yield value can be expressed as

$$\sigma_\beta = \frac{3\varphi_H^2}{2\pi^2 (R+\delta)^2} E_{sep} \qquad (13.31)$$

where ϕ_H is the hydrodynamic volume fraction of the particles that is equal to the effective volume fraction

$$\varphi_H = \varphi_s \left[1 + \frac{\delta}{R}\right]^3 \qquad (13.32)$$

$(R + \delta)$ is the interaction radius of the particle, and E_{sep} is the energy needed to separate a doublet, which is the sum of van der Waals and steric attractions:

$$E_{sep} = \frac{AR}{12H_0} + G_s \qquad (13.33)$$

At a particle separation of ~12 nm (twice the grafted polymer layer thickness), the van der Waals attraction is very small (1.66 kT, where k is the Boltzmann constant and T is the absolute temperature) and the contribution of G_s to the attraction is significantly larger than the van der Waals attraction. Therefore, E_{sep} may be approximated to G_s.

From Eq. (13.33), one can estimate E_{sep} from σ_β. The results are shown in Table 13.4 which show an increase in E_{sep} with increase in ϕ_s.

The values of E_{sep} given in Table 13.4 are unrealistically high, and hence the assumptions made for calculation of E_{sep} are not fully justified and hence the data of Table 13.4 must only be considered as qualitative.

Table 13.4 Results of E_{sep} calculated from σ_β for a flocculated sterically stabilized latex dispersions at various latex volume fractions.

0.4 mol dm^{-3} Na$_2$SO$_4$

ϕ_s	σ_β (Pa)	E_{sep}/kT
0.43	1.3	97
0.45	2.4	165
0.51	3.3	179
0.54	5.3	262
0.55	7.3	336
0.57	9.1	397
0.58	17.4	736
0.25	3.5	804
0.29	5.4	910
0.33	7.4	961
0.37	11.4	1170
0.41	14.1	1190
0.44	17.0	1240
0.47	21.1	1380
0.49	23.1	1390
0.52	28.3	1510

13.1.8.2 Elastic Floc Model

This model [42–44] was described before, and it is based on the assumption that the structural units are small flocs of particles (called flocculi) which are characterized by the extent to which the structure is able to entrap the dispersion medium. A floc is made from an aggregate of several flocculi. The latter may range from loose open structure (if the attractive forces between the particles are strong) to very close-packed structure with little entrapped liquid (if the attractive forces are weak). In the system of flocculated sterically stabilized dispersions, the structure of the floicculi depends on the volume fraction of the solid and how far the system is from the critical flocculation concentration (CFC). Just above the CFC, the flocculi are probably closely packed (with relatively small floc volume), whereas far above the CFC a more open structure is found which entraps a considerable amount of liquid. Both types of flocculi persist at high shear rates, although the flocculi with weak attraction may become more compact by maximizing the number of interactions within the flocculus.

As discussed before, the Bingham yield value is given by Eq. (13.30) which allows one to obtain the floc radius a_{floc} provided one can calculate the floc volume ratio C_{FP} (ϕ_F/ϕ_s) and assumes a value for Δ (the distance through which bonds are stretched inside the floc by the shearing force.

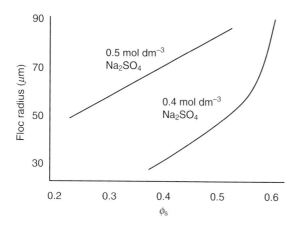

Figure 13.23 Floc radius (a_{floc}) as a function of latex volume fraction at 0.4 and 0.5 mol dm^{-3} (above the CFC).

At high volume fractions, ϕ_F and hence C_{FP} can be calculated using the Krieger equation [3]:

$$\eta_{pl} = \eta_0 \left(1 - \frac{\phi_F}{\phi_s^m}\right)^{-[\eta]\phi_s^m} \tag{13.34}$$

where η_0 is the viscosity of the medium, ϕ_s^m is the maximum packing fraction which may be taken as 0.74, and $[\eta]$ is the intrinsic viscosity taken as 2.5.

Assuming a value of Δ of 0.5 nm, the floc radius R_{floc} was calculatd using Eq. (13.30). Figure 13.23 shows the variation of R_{floc} with latex volume fraction at the two Na$_2$SO$_4$ concentrations studied. At any given electrolyte concentration, the floc radius increases with increase of the latex volume fraction, as expected. This can be understood by assuming that the larger flocs are formed by fusion of two flocs and the smaller flocs by "splitting" of the larger ones. From simple statistical arguments, one can predict that a_{floc} will increase with increase in ϕ_s because in this case larger flocs are favored over smaller ones. In addition, at any given volume fraction of latex, the floc radius increases with increase in electrolyte concentration. This is consistent with the scaling results as discussed above.

The above results show clearly the correlation of viscoelasticity of flocculated dispersions with their interparticle attraction. These measurements allow one to obtain the CFC and CFT of concentrated flocculated dispersions with reasonable accuracy. In addition, the results obtained can be analyzed using various models to obtain some characteristics of the flocculated structure, such as the "openness" of the network, the liquid entrapped in the floc structure, and the floc radius. Clearly, several assumptions have to be made, but the trends obtained are consistent with expectation from theory.

References

1 Einstein, A. (1906) *Ann. Phys.*, **19**, 289; Einstein, A. (1911) *Ann. Phys.*, **34**, 591.
2 Batchelor, G.K. (1977) *J. Fluid Mech.*, **83**, 97.
3 Krieger, I.M., and Dougherty, T.J. (1959) *Trans. Soc. Rheol.*, **3**, 137.
4 Krieger, I.M. (1972) *Adv. Colloid Interface Sci.*, **3**, 111.
5 Goodwin, J.W., and Hughes, R.W. (2000) *Rheology for Chemists*, Royal Society of Chemistry Publication, Cambridge.
6 Tadros, T.F. (1989) Rheology of concentrated stable and flocculated suspension, in *Flocculation and Dewatering* (eds B.M. Mougdil and B.J. Scheiter), Engineering Foundation Publishers, USA, pp. 43–87.
7 Tadros, T.F. (1996) *Adv. Colloid Interface Sci.*, **68**, 97.
8 Tadros, T.F. (1990) *Langmuir*, **6**, 28.
9 Tadros, T.F., and Hopkinson, A. (1990) *Faraday Discuss. Chem. Soc.*, **90**, 41.
10 Napper, D.H. (1983) *Polymeric Stabilization of Colloidal Dispersions*, Academic Press, London.
11 Liang, W., Tadros, T.F., and Luckham, P.F. (1992) *J. Colloid Interface Sci.*, **153**, 131.
12 Prestidge, C., and Tadros, T.F. (1988) *J. Colloid Interface Sci.*, **124**, 660.
13 de L. Costello, B.A., Luckham, P.F., and Tadros, T.F. (1988/1989) *Colloids Surf.*, **34**, 301.
14 Luckham, P.F., Ansarifar, M.A., de L. Costello, B.A., and Tadros, T.F. (1991) *Powder Technol.*, **65**, 371.
15 Tadros, T.F., Liang, W., de L. Costello, B.A., and Luckham, P.F. (1993) *Colloids Surf.*, **79**, 105.
16 Luckham, P.F. (1989) *Powder Technol.*, **58**, 75.
17 de Gennes, P.G. (1987) *Adv. Colloid Interface Sci.*, **27**, 189.
18 Russel, W.B. (1980) *J. Rheol.*, **24**, 287.
19 Hoffman, R.L. (1983) *Science and Technology of Polymer Colloids*, vol. II (eds G.W. Poehlein, R.H. Ottewill, and J.W. Goodwin), Martinus Nijhoff Publishers, Boston, The Hague, p. 570.
20 Firth, B.A., and Hunter, R.J. (1976) *J. Colloid Interface Sci.*, **57**, 248.
21 van de Ven, T.G.M., and Hunter, R.J. (1976) *Rheol. Acta*, **16**, 534.
22 Hunter, R.J., and Frayane, J. (1980) *J. Colloid Interface Sci.*, **76**, 107.
23 Heath, D., and Tadros, T.F. (1983) *Faraday Discuss. Chem. Soc.*, **76**, 203.
24 Prestidge, C., and Tadros, T.F. (1988) *Colloids Surf.*, **31**, 325.
25 Tadros, T.F., and Zsednai, A. (1990) *Colloids Surf.*, **49**, 103.
26 Liang, W., Tadros, T.F., and Luckham, P.F. (1993) *J. Colloid Interface Sci.*, **155**, 156.
27 Liang, W., Tadros, T.F., and Luckham, P.F. (1993) *J. Colloid Interface Sci.*, **160**, 183.
28 Ball, R., and Brown, W.D. (1990) personal communication.
29 Buscall, R., and Mill, P.D.A. (1988) *J. Chem. Soc. Faraday Trans. I*, **84**, 4249.
30 Gast, A.P., Hall, C.K., and Russel, W.B. (1983) *J. Colloid Interface Sci.*, **96**, 251.
31 Asakura, S., and Oosawa, F. (1954) *J. Chem. Phys.*, **22**, 1255.
32 Asakura, S., and Oosawa, F. (1958) *J. Polym. Sci.*, **33**, 183.
33 Fleer, G.J., Scheutjens, J.H.M.H., and Vincent, B. (1984) *ACS Symp. Ser.*, **240**, 245.
34 Goodeve, C.V. (1939) *Trans. Faraday Soc.*, **35**, 342.
35 Gillespie, T. (1960) *J. Colloid Sci.*, **15**, 219.
36 Hunter, R.J., and Nicol, S.K. (1968) *J. Colloid Interface Sci.*, **28**, 200.
37 Firth, B.A., and Hunter, R.J. (1976) *J. Colloid Interface Sci.*, **57**, 248, 257, 266.
38 Mills, P.D.A., Goodwin, J.W., and Grover, B. (1991) *Colloid Polym. Sci.*, **269**, 949.
39 Goodwin, J.W., and Hughes, R.W. (1992) *Adv. Colloid Interface Sci.*, **42**, 303.
40 Liang, W., Tadros, T.F., and Luckham, P.F. (1983) *Langmuir*, **9**, 2077.
41 Firth, B.A., Neville, P.C., and Hunter, R.J. (1974) *J. Colloid Interface Sci.*, **49**, 214.
42 van de Ven, T.G.M., and Hunter, R.J. (1974) *J. Colloid Interface Sci.*, **68**, 135.
43 Hunter, R.J. (1982) *Adv. Colloid Interface Sci.*, **17**, 197.
44 Friend, J.P., and Hunter, R.J. (1971) *J. Colloid Interface Sci.*, **37**, 548.

Index

a
AcoustoSizer cell 79
adhesion
– tension 20–22
– work of 22
adsorption
– adsorbate–adsorbent interaction 100
– adsorbed layer thickness 119–122, 126–128
– dynamic processes 34–42
– energy 109
– Gibbs equation 5, 33
– isotherm, see isotherm
– kinetics 34–42, 128
– measurements 196–199
– (non)specific adsorbed ions 52–53
– polymeric surfactants 107–129, 131–136
– polymers 110–117
– relative 3
– surfactants 33–34, 93–103
– temperature effects 104
advancing angle 26
Aerosol OT 46
AFM (atomic force microscopy) 206
agents
– dispersing 2, 85–105
– wetting 45–46
aggregation
– aggregate/agglomerate breaking 8
– configurational entropy 167
– diffusion-controlled 254
– rapid/slow 260
air/solution interface 103
alcohol ethoxylates 87–88
alkyl–alkyl interaction 101
alkyl naphthalene formaldehyde condensates 93
alkyl phenol ethoxylates 88
alumina substrates 98

amine ethoxylates 90
amphoteric surfactants 86–87
analysis of creep curves 226
analysis of flow curves 220–223, 258–259
anchor 125–127
angle, contact 18–20, 23–28
anionic surfactants 85–86
antithixotropy 223–225
aqueous clay dispersions 181
aqueous media 179
assessment
– flocculation 201–206
– sedimentation 199–201
– stability, see stability assessment
athermal solvent 115
Atlox 4913 147
atom arrangement 180
atomic force microscopy (AFM) 206
autocorrelation function 212

b
backscattering techniques 214
balance of density 178
Batchelor equation 244
benzyl alcohol 151
Bingham plastic systems 221
Bingham yield value 251, 254, 264
block copolymers 108, 124
bonding, hydrogen 139
Born repulsion 57
Brag–Williams approximation 112
breaking, aggregates/agglomerates 8
bridging flocculation 138–141, 165
Brownian diffusion 160, 166
– concentrated suspensions 243–245
– energy 171
Brownian motion 121
bubble pressure 39–42
bulk properties, suspensions 216

Dispersion of Powders in Liquids and Stabilization of Suspensions, First Edition. Tharwat F. Tadros.
© 2012 Wiley-VCH Verlag GmbH & Co. KGaA. Published 2012 by Wiley-VCH Verlag GmbH & Co. KGaA.

Bulkley, Herschel–Bulkley equation   222
buoy   125–127

c

calculation
– contact angle   23–25
– surface tension   23–25
– zeta potential   68–73
capillary flow   40
capture distance   152
carbon substrates   96
carboxylates   86
Casson model   222
cationic polyelectrolytes   141
cationic surfactants   86, 240
cell
– AcoustoSizer   79
– cylindrical electrophoretic   74
cellulose
– EHEC   176–177
– hydroxyethyl   11, 190
cetyl trimethyl ammonium bromide (CTAB)   240
CFC (critical flocculation concentration)   263–264, 268–269
CFT (critical flocculation temperature)   215, 241
CFV (critical volume fraction)   138
characterization, suspensions   193–217
charge
– accumulation   64
– density   194
– separation   64
– surface   49–51, 194–195
chemical potential   256
– solvents   133
claying   191
clays   10–11, 51
– aqueous dispersions   181
– restabilization   240
close-packed hemimicelles   101
CLSM (confocal laser scanning microscopy)   205
coagulation   156, 165
– electrolytes   261–263
– suspensions   258, 261–263
coalescence   236–237
coarse suspensions   165
cohesion, work of   22–23
cohesive energy density   232–233
colloidal dispersions   162, 165
colloidal vibration potential (CVP)   79
comminution   8–9
compact sediments   10–11

complete wetting   19
complex modulus   230–231
compression
– liquids   35
– particles   132
concentrated suspensions   12, 151–169
– rheology   243–270
concentration
– CFC   263–264, 268–269
– cmc   87, 97–102
– electrolytes   157–158, 185
– particles   171
– range   11–12, 82
– surfactants   2
condensation methods   142
configurational entropy   109, 134, 154–155
– particle aggregation   167
confocal laser scanning microscopy (CLSM)   205
conformation, polymeric surfactants   107–129
constant stress measurements   176, 219, 225–229
contact angle   18–20
– calculation   23–25
– hysteresis   26–28
– measurement   44–45
– water drops   18
controlled flocculation   183–186
copolymers   142–148
– block   108, 124
– diblock   125–127
– graft   124
counterions, valency   152
coverage, surface   116
creaming   234
creep curves, analysis   226
creep measurements   176, 219, 225–229
critical flocculation concentration (CFC)   263–264, 268–269
critical flocculation temperature (CFT)   215, 241
critical micelle concentration (cmc)   87, 97–102
critical surface tension   31–47
critical volume fraction (CFV)   138
Cross equation   222–223
crystal growth   9–10, 201–206
– measurement   216
CTAB (cetyl trimethyl ammonium bromide)   240
cumulant method   213
CVP (colloidal vibration potential)   79
cylindrical electrophoretic cell   74

d

Deborah number   243
Debye, RGD regime   207–208
Debye–Hückel parameter   66–67
decay constant   212
density
– balance   178
– charge   194
– cohesive energy   232–233
– segment density distribution   111–112, 119–122
depletion flocculation   138, 165–168
– free energy   255
– osmotic pressure   168
– sedimentation   186–190
depth of immersion   27
Deryaguin–Landau–Verwey–Overbeek theory   57–59, 157, 183–184
desorption flux   35
destabilization, suspensions   131–149
Deuhem, Gibbs–Deuhem equation   3
deuterated polystyrene latex   126–127
dewetting   27
diblock copolymers   125–127
diethylhexyl sulphosuccinate   46
differential interference contrast   202–203
diffraction techniques, light   208–210
diffuse double layer   51–52
diffusion
– Brownian   160, 166, 171, 243–245
– diffusion-controlled aggregation   254
– Stokes–Einstein equation   60
– translational coefficient   212
dilatant sediments   178
dilatant systems   222
dilation, liquids   35
dilute region   114
dilute solutions   175
dilute suspensions   12, 160–164
dispersing agents   2, 85–105
dispersion constant, London   23, 54, 156
dispersion methods   142
dispersion polymerization   142
dispersions
– colloidal   162, 165
– microstructure   160
– NAD   142
– polystyrene   174
– polystyrene latex   164, 247–248
– powders   1–11
– preformed latex   146–148
– stabilization   9
– steric stabilization   136–138, 240–241, 248–250

distance
– atoms   180
– capture   152
– energy–distance curves   135–137, 158–159, 184
– force–separation distance curves   144–145
distribution function, radial   161
divided particulate solids   182–183
DLVO theory   57–59, 157, 183–184
Doppler shift   78
dorn effect   65
double layer
– characterization   194–196
– double-layer interaction   53
– double-layer overlap   181
– electrical   51–54, 67–68
– thickness   67
doublet floc structure model   267–268
Dougherty–Krieger equation   173, 246–248
drop profile   27–28
drop volume technique   38–39
dynamic adsorption and wetting processes   34–42
dynamic light scattering   211–214
dynamic mobility   80
dynamic (oscillatory) measurements   220, 229–241
dynamic viscosity   230–231

e

edge-to-face association   181
effective Hamaker constant   55–56
effective steric stabilization   135–136
EHEC (ethyl(hydroxyethyl) cellulose)   176–177
Einstein
– Einstein equation   120, 244
– Stokes–Einstein equation   60, 77, 121
elastic floc model   259, 268–269
elastic interaction   134–135
elastic modulus   260
elastic response   151, 225
electroosmotic velocity   69
electrical double layer   51–54, 67–68
– repulsion   53–54
electroacoustic methods   78
electrokinetic phenomena   63–83
– measurements   195–196
electrolytes   54, 59
– coagulation   261–263
– concentration   157–158, 185
– electroacoustic potential   82
electron microscopy   204–205

electroosmosis   64–65
electrophoresis   64
– cylindrical cell   74
– electrophoretic mobility   73–78, 121, 240
electrostatic interaction   152–153, 246–248
– electrostatic-patch model   141
electrostatic stabilization   49–62
– emulsions   61
– suspensions   184, 261–263
ellipsometry   119
emulsion polymerization   142
energy
– adsorption   109
– Brownian diffusion   171
– cohesive   232–233
– energy–distance curves   135–137, 158–159, 184
– free   186, 255
entropy, configurational   109, 134, 154–155, 167
equilibrium sediment volume/height   216–217
esters, sorbitan   89–90
ethoxylated fats and oils   90
ethyl(hydroxyethyl) cellulose (EHEC)   176–177
experimental techniques
– adsorption kinetics   37–42
– polymeric surfactant adsorption   117–118

f
fatty acid ethoxylates   88
fine particles, "inert"   179–182
finely divided particulate solids   182–183
flocculation   59–62, 165
– assessment   201–206
– bridging   138–141, 165
– CFC   263–264, 268–269
– clays   240
– concentrated suspensions   245
– controlled   183–186
– critical flocculation temperature   215, 241
– depletion   186–190
– doublet floc structure model   267–268
– elastic floc model   259, 268–269
– flocculated suspension rheology   250–258
– fractal concept   259–260
– free energy   186
– rate measurements   214–215
– rheological measurements   235–241
– sterically stabilized dispersions   136–138, 240–241
– temperature effect   266–267
Flory–Huggins interaction parameter   113, 132, 154–155

flow
– capillary   40
– flow curve analysis   220–223, 258–259
– Marangoni   35
– pseudoplastic flow curve   258
– thixotropy   223–225
fluids
– "fluid-like" behavior   151
– Newtonian   221
– non-Newtonian   173
force
– force–separation distance curves   144–145
– gravity   172
– hydrodynamic   68, 172
– van der Waals   8, 23, 267
Fowkes treatment   25
fractal flocculation concept   259–260
fraction of segments   118–119
Fraunhofer theory   209–210
free energy   186, 255
free polymers   252–253
frequency sweep   231
friction, hydrodynamic   82
Frumkin–Fowler–Guggenheim equation   95

g
Gans, RGD regime   207–208
gels   179
general scaling law   257
Gibbs adsorption equation   5, 33
Gibbs–Deuhem equation   3
"girders"   259
glass   31
– hydrophobized surfaces   145
glycerol   91
goethite   195–196
Good and Girifalco approach   24–25
graft copolymers   124
Grahame, Stern–Grahame model   52, 67–68
graphitization   97
gravity force   172
gum, xanthan   11
gyration, radius   168, 256

h
Hamaker constant   23, 156
– effective   55–56
hard-sphere interaction   151–152
hard-sphere suspensions   244
head group, hydrophilic   4
HEC (hydroxyethyl cellulose)   11, 190

height, sediment   185, 216–217
hemimicelles, close-packed   101
Henry region, linear   115
Henry's treatment   72–73
Herschel–Bulkley equation   222
hexagonal packing   121
high-affinity isotherm   199
high-energy solids   31
high-frequency shear modulus   164
high molecular weight, thickeners   178–179
HLB (hydrophilic-lipophilic-balance)   99
homopolymer sequence   108
Hückel equation   71–72
– Debye–Hückel parameter   66–67
Huggins, Flory–Huggins interaction parameter   113, 132, 154–155
hydrodynamic force   68, 172
hydrodynamic friction   82
hydrodynamic thickness   127, 131
hydrogen bonding   139
hydrophilic head group   4
hydrophilic-lipophilic-balance (HLB)   99
hydrophobic powders   1
hydrophobic solids   45–46, 103–104
hydrophobic surfaces   94–97
hydrophobically modified inulin copolymers   142–146
hydrophobized glass surfaces   145
hydroxyethyl cellulose (HEC)   11, 190
hypermers   146–147

i
immersion
– depth of   27
– solids   21
– test   45
impulse theory   258–259
incipient flocculation   137–138
– measurement   215–216
incomplete wetting   19
"inert" fine particles   179–182
insulin backbone   93
intensity fluctuation   76, 211
interaction
– adsorbate–adsorbent   100
– alkyl–alkyl   101
– double-layer   53
– elastic   134–135
– electrostatic   152–153, 246–248
– Flory–Huggins   113, 132, 154–155
– hard-sphere   151–152
– interparticle   151–159
– mixing   132–134
– soft   152–153, 244

– steric   131–136, 153–156
– van der Waals, *see* van der Waals forces
interfaces
– air/solution   103
– charge accumulation   64
– solid/liquid   49–62, 93–103, 107–129, 194–199
interference contrast, differential   202–203
interpenetration   132
INUTEC®   92, 142–146
ionic surfactants   5, 94–98
ionic vibration potential (IVP)   78–79
ions
– adsorbed   52–53
– surface   49–50
isoelectric point   98
isomorphic substitution   50–51
isotherm   103–104
– high affinity   199
– Langmuir-type   6, 197
– measurement   118
– nonionic polymeric surfactants   122–126
– Stern–Langmuir   94

j
Jones, Lennard–Jones equation   24

k
Kelvin equation   216
kinetics
– adsorption   34–42
– flocculation   59–60
– micellar   37–38
– polymer adsorption   128
Krieger
– Dougherty–Krieger equation   173, 246–248
– Krieger equation   269

l
Landau, Deryaguin–Landau–Verwey–Overbeek theory   57–59, 157, 183–184
Langmuir
– Langmuir-type adsorption isotherm   6, 197
– Stern–Langmuir equation   6–7
– Stern–Langmuir isotherm   94
Laplace transform   213
laser scanning microscopy, confocal   205
laser velocimetry   76–78
latex
– deuterated polystyrene   126–127
– polystyrene latex sediments   187
– preformed dispersions   146–148

lattice
- quasicrystalline   113
- standard   210

laws and equations
- Batchelor equation   244
- Cross equation   222–223
- Debye–Hückel parameter   66–67
- Dougherty–Krieger equation   173, 246
- dynamic viscosity   230–231
- Einstein equation   120, 244
- Frumkin–Fowler–Guggenheim equation   95
- general scaling law   257
- Gibbs adsorption equation   5, 33
- Gibbs–Deuhem equation   3
- Herschel–Bulkley equation   222
- Hückel equation   71–72
- Kelvin equation   216
- Krieger equation   269
- Lennard–Jones equation   24
- Nernst equation   194
- Ohm's law   70
- osmotic pressure   163
- particle concentration   171
- photocount correlation function   77
- Poiseuille's law   41
- Poisson's equation   69
- Rideal–Washburn equation   43–44
- shear stress   176
- Stern–Langmuir equation   6–7
- Stokes–Einstein equation   60, 77, 121
- Stokes' law   172
- Wenzel's equation   28–29
- Young's equation   19–20

layer overlap   133
Lennard–Jones equation   24
light diffraction techniques   208–210
light scattering techniques   76, 207–208
linear Henry region   115
liquid crystalline phases   190–191
liquids
- compression/dilation   35
- contact angle measurement   44–45
- effective Hamaker constant   56
- Rideal–Washburn equation   43–44
- solid/liquid interface   49–62, 93–103, 107–129, 194–199
- supernatant   73
- surface spreading   25–26
- viscoelastic   226
- viscosity   65–66

log W–log C curve   61
London dispersion constant   23, 54, 156

loops   108, 117
loss modulus   266
low-energy solids   31

m

macromolecules   108
macromonomers   146
Marangoni flow   35
maximum bubble pressure technique   39–42
Maxwell relaxation time   230
mean-field approximation   112
measurements   37–42, 117–118
- adsorption isotherm   118
- adsorption kinetics   37–42
- constant stress   176, 219, 225–229
- contact angle   44–45
- dynamic (oscillatory)   220, 229–241
- electrokinetic   195–196
- electrophoretic mobility   73–78
- flocculation rate   214–215
- incipient flocculation   215–216
- particle size   203
- rheological   188, 219–241
- steady-state   220–225
- surfactant/polymer adsorption   196–199
- turbidity   208
- see also assessment, experimental techniques

methods and techniques   21
- cumulant method   213
- electroacoustic   78
- Fowkes treatment   25
- Good and Girifalco approach   24–25
- Henry's treatment   72–73
- Monte Carlo   111
- random walk approach   110
- scattering techniques   206–214
- sedimentation method   121
- viscosity method   121
- von Smoluchowski treatment   68–71

micellar solutions   36–37
microelectrophoresis   73–76
microscopy
- electron   204–205
- optical   201–203
- ultramicroscopic technique   73–76, 202

microstructure, dispersion   160
Mie regime   208
milling, wet   8–9
mixing interaction   132–134
mixtures of polymers   182–183

mobility
- dynamic 80
- electrophoretic 73–78, 121, 240
- particle 68–73
model nonionic surfactants 103–104
modulus
- complex 230–231
- elastic 260
- high-frequency shear 164
- loss 266
- storage 257, 266
molecular weight 178–179
Monte Carlo method 111
montmorillonite, sodium 179, 182

n

Na, see sodium
negative thixotropy 223–225
Nernst equation 194
network, three-dimensional 250
(non-)Newtonian fluids 173, 220–221
nonabsorbing particles 208
nonaqueous dispersion polymerization (NAD) 142
nonionic surfactants 87, 98–101
- model 103–104
- polymeric 122–126
nonspecific adsorbed ions 52–53

o

Ohm's law 70
oils, ethoxylated 90
oligomers, dilute region 114
optical microscopy 201–203
oscillatory measurements 220, 229–241
oscillatory sweep 232, 239–240
osmotic pressure 139, 163
- depletion flocculation 168
Ostwald ripening 9–10, 201–206
- measurement 216, 236–237
Overbeek, Deryaguin–Landau–Verwey–Overbeek theory 57–59, 157, 183–184
overlap 133, 181
oxide surface 49–50

p

pair distribution function 161–162
partial wetting 19
particles
- aggregation 167
- concentration 82, 171
- dimensions 11
- "inert" fine 179–182
- mobility 68–73
- nonabsorbing 208
- size 178, 203
- spherical 153
- steric stabilization 131–136
particulate solids 182–183
Péclet number 243
penetration, rate of 43–44
PEO (poly(ethylene oxide)) 91, 99, 108, 122–123
- sedimentation 188–190
- stability assessment 198
permittivity 81, 152
phase contrast 202
phosphates 86, 191
photocount correlation function 77
photon correlation spectroscopy (PCS) 211–214
plate technique, Wilhelmy 21
Poiseuille's law 41
Poisson's equation 69
polar surfaces 97–98
polarized light microscopy 203
polydisperse suspensions 213, 247
polyelectrolytes 138–141
- cationic 141
poly(ethylene oxide) (PEO) 91, 99, 108, 122–123
- sedimentation 188–190
- stability assessment 198
polyethylene terephthalate 31
polyfructose 143
polymeric surfactants 90–93
- adsorption isotherm 122–126
- conformation 107–129
- nonionic 122–126
- steric stabilization 131–149
polymerization 142
polymers
- adsorption 110–117, 128, 196–199
- bridging flocculation 138–141
- dilute region 114
- free 252–253
- layer overlap 133
polystyrene dispersions 174
polystyrene latex
- deuterated 126–127
- dispersions 164, 247–248
- sediments 187
polystyrene latex suspensions, coagulated 262–263
polytetrafluoroethylene (PTFE) 31–32

poly(vinyl alcohol) (PVA)   107, 123–124
– sedimentation   186
porous substrates   17
powders
– contact angle measurement   44–45
– dispersion   1–11
– hydrophobic   1
– spreading   17–29
– wettability   45
– wetting, *see* wetting
power law plot   254
precipitated silica   194–195
preformed latex dispersions   146–148
pressure
– bubble   39–42
– osmotic   139, 163, 168
pseudoplastic flow curve   258
pseudoplastic systems   221
PTFE (polytetrafluoroethylene)   31–32
PVA (poly(vinyl alcohol))   107, 123–124
– sedimentation   186
pzc value   50

q
quasicrystalline lattice   113

r
radial distribution function   161
radius of gyration   168, 256
random walk approach   110
rapid aggregation   260
rate of penetration   43–44
Rayleigh regime   207–208
receding angle   26
redispersion   216–217
– sediment   185
relative adsorption   3
relative viscosity   246
relaxation time   239
– Maxwell   230
residual viscosity   228, 234
restabilization   140
– clays   240
reversible time dependence, viscosity   179
rheological measurements   188, 219–241
rheology, concentrated suspensions   243–270
Rideal–Washburn equation   43–44
ripening, Ostwald   9–10, 201–206, 216, 236–237
rough surfaces   28

s
sample preparation, optical microscopy   203
scaling law   257

scanning electron microscopy (SEM)   204–205
scanning probe microscopy (SPM)   205
scanning tunneling microscopy (STM)   206
scattered light   76
scattering techniques   206–214
sedimentation   10–11
– assessment   199–201
– dilatant sediments   178
– polystyrene latex sediments   187
– potential   65
– rate   172–178, 234
– rheological measurements   233–235
– sediment height   185, 216–217
– suspensions   171–192
– velocity   172
sedimentation method   121
segments
– density distribution   111–112, 119–122
– fraction of   118–119
"self-structured" systems   183–186
semidilute solutions   175
separation, charge   64
shear
– high-frequency modulus   164
– plane   63
– stress   176–177, 219
– thinning/thickening systems   221
silica   179
– precipitated   194–195
silicates   191
sinking time   45
size reduction, particles   178
slow aggregation   260
Smoluchowski rate   215
Smoluchowski treatment   68–71
sodium dodecyl sulfate   96–98
sodium montmorillonite   179, 182
soft interaction   244
– electrostatic   152–153, 246–248
"solid-like" behavior   151
solid/liquid interface   49–62, 93–103
– characterization   194–199
– polymeric surfactants   107–129
solid substrates   22
solid suspensions   160–164
solids
– hydrophobic   45–46, 103–104
– immersion   21
– low-/high-energy   31
– particulate   182–183
– viscoelastic   227

solutions
– air/solution interface   103
– athermal solvents   115
– micellar   36–37
– reduced solvency   137
– (semi)dilute   175
– solvent chemical potential   133
– surfactant   44–45
sorbitan esters   89–90
spans   89–90
specific adsorbed ions   52–53
spherical particles   153
SPM (scanning probe microscopy)   205
spontaneous spreading   18
spontaneous wetting   44
spoon   238
spreading   17–29
– spontaneous   18
– surfaces   25–26
stability assessment   193–217
– rheological   219–241
stability ratio   215
stabilization
– dispersions   9
– electrostatic   49–62
– emulsions   61
– suspensions   131–149
standard lattices   210
standing suspensions   164–168
steady-state measurements   220–225
steady-state shear stress   219
step change   225
steric interaction   153–156
steric stabilization   131–149
– dispersions   136–138, 240–241, 248–250
– suspensions   245
Stern–Grahame model   52, 67–68
Stern–Langmuir equation   6–7
Stern–Langmuir isotherm   94
STM (scanning tunneling microscopy)   206
Stokes–Einstein equation   60, 77, 121
Stokes' law   172
Stokes' velocity   172
storage modulus   257, 266
strain sweep   231–232, 238–239
– latex dispersions   261
streaming potential   65
stress
– shear   176–177, 219
– thermal   243
– yield   260
strong flocculation   258, 261–263

structure
– Aerosol OT   46
– amine ethoxylates   90
– assessment   194–199
– clays   51
– doublet floc structure model   267–268
– electrical double layer   51–52
– INUTEC®   92
– "self-structured" systems   183–186
– solid/liquid interface   49–62
submersion   45
substitution, isomorphic   50–51
substrates
– alumina   98
– carbon   96
– porous   17
– solid   22
sulfates   86
sulfonated alkyl naphthalene formaldehyde condensates   93
sulfonates   86
supernatant liquids   73
surfaces
– charge   49–51, 194–195
– coverage   116
– critical tension   31–47
– hydrophobic   94–97, 145
– ions   49–51
– liquid spreading   25–26
– oxide   49–50
– polar   97–98
– rough   28
– tension   2, 23–25, 36
– tilted   27
surfactants
– adsorption   33–34, 93–103, 196–199
– amphoteric (zwitterionic)   86–87
– anionic   85–86
– cationic   86, 240
– concentration   2
– contact angle measurement   44–45
– ionic   5, 94–98
– nonionic   87–88, 98–104
– polymeric   90–93, 107–129
– steric stabilization   131–136
suspensions
– bulk properties   216
– characterization   193–217
– coagulated   258, 261–263
– coarse   165
– concentrated   151–169, 243–270
– concentration range   11–12
– (de)stabilization   131–149
– dilute/concentrated   12, 160–164, 243–270

– electrostatically stabilized   184, 261–263
– flocculated   59–62, 250–258
– particle dimensions   11
– polydisperse   213, 247
– rheological measurements   219–241
– rheology   243–270
– sedimentation   171–192
– solid   160–164
– standing   164–168
– sterically stabilized   245
– viscoelastic properties   249–250
syneresis   233–235

t
tails   117
TEM (transmission electron microscopy)   204
temperature
– CFT   215, 241
– effect on adsorption   104
– effect on flocculation   266–267
– electroacoustic methods   82–83
tension
– adhesion   20–22
– critical   31–47
– surface   2, 23–25, 36
– wetting   20
tetrafunctional products   91
theories and models
– Casson model   222
– DLVO theory   57–59, 157, 183–184
– doublet floc structure model   267–268
– elastic floc model   259, 268–269
– electrostatic-patch model   141
– fractal flocculation concept   259–260
– Fraunhofer theory   209–210
– impulse theory   258–259
– model nonionic surfactants   103–104
– Stern–Grahame model   52, 67–68
thermal stress   243
thermodynamic treatment   19–20
theta solvent   115
thickeners   235
– high molecular weight   178–179
thickness
– adsorbed layer   119–122, 126–128
– hydrodynamic   127, 131
thixotropy   179
– flow   223–225
– negative   223–225
three-dimensional network   250
three-phase line   19
tilted surfaces   27
time-average light scattering   207–208

time dependence, viscosity   179
translational diffusion coefficient   212
transmission electron microscopy (TEM)   204
trifunctional products   91
tristyrylphenol   88
turbidity measurements   208
tweens   89–90

u
ultramicroscopic technique   73–76, 202
ultrasound vibration potential (UVP)   79

v
valency, counterions   152
van der Waals forces   8, 23, 54–57, 156–157
– doublet floc structure model   267
vector analysis, complex modulus   230–231
velocity
– electroosmotic   69
– laser velocimetry   76–78
– sedimentation   172
Verwey, Deryaguin–Landau–Verwey–Overbeek theory   57–59, 157, 183–184
viscoelastic systems   225
– oscillatory response   229–230
– suspensions   249–250
viscosity   65–66
– dynamic   230–231
– relative   246
– residual   228, 234
– reversible time dependence   179
– viscosity–volume fraction curve   246
– viscous response   151, 225
– zero shear   228, 234
viscosity method   121
von Smoluchowski, see Smoluchowski

w
Washburn, Rideal–Washburn equation   43–44
water drops, contact angle   18
weak flocculation   136–137, 165, 251–258
Wenzel's equation   28–29
wet milling   8–9
wettability, powders   45
wetting   1–8, 17–29
– critical surface tension   31–47
– dynamic processes   34–42
– spontaneous   44
– surfactants   31–47
– tension   20
– wetting agents   45–46
– wetting line   19

Wilhelmy plate technique   21
– methods and techniques   21
work, ad-/cohesion   22–23

x
X-ray sedimentation   200
xanthan gum   11

y
yield stress   260
yield value, Bingham   251, 254, 264
yield variation   182, 189, 240
Young's equation   19–20

z
zero shear viscosity   228, 234
zeta potential   63–83
– calculation   68–73
– measurement   73–78, 195–196
zwitterionic surfactants   86–87